DEVRY INSTITUTE OF TECHNOLOGY
LIBRARY
MISSISSAUGA, ONTARIO

Analog-to-Digital Conversion

DEVRY INSTITUTE OF TECHNOLOGY
LIBRARY
MISSISSAUGA, ONTARIO

Analog-to-Digital Conversion

A Practical Approach

Kevin M. Daugherty

McGraw-Hill, Inc.

New York San Francisco Washington, D.C. Auckland Bogotá
Caracas Lisbon London Madrid Mexico City Milan
Montreal New Delhi San Juan Singapore
Sydney Tokyo Toronto

Library of Congress Cataloging-in-Publication Data

Daugherty, Kevin M.
Analog-to-digital conversion : a practical approach / Kevin M. Daugherty.
p. cm.
Includes bibliographical references and index.
ISBN 0-07-015675-1
1. Analog-to-digital converters. I. Title.
TK7887.6.D38 1994
621.39'814—dc20 94-5266
CIP

Copyright © 1995 by McGraw-Hill, Inc. All rights reserved. Printed in the United States of America. Except as permitted under the United States Copyright Act of 1976, no part of this publication may be reproduced or distributed in any form or by any means, or stored in a database or retrieval system, without the prior written permission of the publisher.

1 2 3 4 5 6 7 8 9 0 DOC/DOC 9 0 9 8 7 6 5 4

ISBN 0-07-015675-1

The sponsoring editor for this book was Stephen S. Chapman, the editing supervisor was Joseph Bertuna, and the production supervisor was Pamela A. Pelton. It was set in Century Schoolbook by McGraw-Hill's Professional Book Group composition unit.

Printed and bound by R. R. Donnelley & Sons Company.

Information contained in this work has been obtained by McGraw-Hill, Inc., from sources believed to be reliable. However, neither McGraw-Hill nor its authors guarantees the accuracy or completeness of any information published herein, and neither McGraw-Hill nor its authors shall be responsible for any errors, omissions, or damages arising out of use of this information. This work is published with the understanding that McGraw-Hill and its authors are supplying information, but are not attempting to render engineering or other professional services. If such services are required, the assistance of an appropriate professional should be sought.

In memory of my father Robert Daugherty

DEVRY INSTITUTE OF TECHNOLOGY
LIBRARY
MISSISSAUGA, ONTARIO

Contents

Preface

Analog-to-digital (A/D) conversion plays a major role in many real-world interfaces, and it is surprising that very few references deal with the practical aspects. Most of the published material available today focuses only on the manufacturers' integrated devices and only lightly covers additional support devices. Many fine points must be considered in order to avoid major problems with final system performance. One area that is especially lacking a good reference source is in the low-cost discrete A/D converters for applications that need to be more cost-competitive.

This book covers the subject of analog-to-digital conversion in a practical and broad fashion. Several low-cost techniques as well as the design requirements for fully integrated high-performance solutions are explained and supported with examples. This includes both hardware and software techniques in many cases. Support components and layout practices are also discussed. Armed with this knowledge, the engineer can quickly choose an optimum approach and understand how to make it work.

Digital signal processing (DSP) has become an important technique in data acquisition systems. This is probably more true with the delta-sigma type of A/D converter than with any other. The idea here is to shift the burden from precision analog circuitry to the highly integrated digital circuitry. There are already several delta-sigma devices available today, and their popularity is likely to continue for a long time to come.

Chapter 1 introduces various A/D conversion techniques and discusses system architectural tradeoffs to aid in the selection process. Comparisons are made between several techniques that include attributes such as accuracy, resolution, speed, and cost. An example of an architectural decision is the tradeoff between a higher-resolution A/D converter or the addition of a programmable gain amplifier. Data acquisition terms are then explained to provide an understanding of important specifications that are often encountered.

Chapter 2 discusses the importance of passive support components (resistors and capacitors) necessary in all data acquisition systems. Paying close attention to these components will ensure that the selected A/D device and technique will not have degraded performance.

Chapter 3 covers the many active support devices such as references, op amps, comparators, sample-and-holds, and multiplexers. Where these devices fit, and how to select them will depend on the desired speed, dc accuracy, and voltage-range requirements.

Chapter 4 explains in detail several low-cost microcontroller-based A/D conversion techniques. Included are several unique methods such as an improved pulse-width modulation (PWM) A/D converter and a potentiometer measurement. Many of these techniques have come from my experience in designing circuits for various high-volume customers. The designs necessitated understanding the complete system requirements since the microcontroller usually needs to perform other tasks.

Chapter 5 covers integrating-type A/D converters (single- and dual-slope) which can also be considered low-cost. Although dual-slope converters are widely available, the single-slope converters have been largely neglected. Chapter 5 covers both techniques and explains how to build an extremely low-cost and highly accurate single-slope circuit that can require very little hardware or software.

Chapter 6 focuses on successive-approximation (sampling) A/D converters. Included are discussions of how the external component selections can greatly affect the performance. By understanding the inner workings of these devices, problems can be avoided. In addition, some techniques are presented for increasing the resolution or range of the A/D converter (dual polarity and dithering). Finally, the wide range of accuracy, features, and level of integration solutions available is illustrated.

Chapter 7 deals with high-speed flash and subranging (i.e., half flash) A/D converters. CMOS and bipolar converters are compared along with their potential sources of errors. This is followed by a discussion of how to optimize for performance and cost.

Chapter 8 explains how delta-sigma (oversampling) A/D converters work. Although this technique has been known for nearly 50 years, it has taken recent state-of-the-art integration methods to make it attractive. Included in Chap. 8 is a brief discussion of digital signal processing (DSP). It is the intention of this book to provide an overview of the main concepts and to leave detailed discussions to the many good books on the topic of DSP. Finally, typical uses of delta-sigma converters are presented.

Chapter 9 covers the error sources that can destroy any A/D conversion design if close attention is not paid to several areas. This includes both ac and dc noise, temperature drifts, and how to minimize the errors. Good layout practice that includes the use of guard rings and shielding is also discussed. The last section gives ideas on how to calibrate out errors (i.e., autozeroing).

Kevin M. Daugherty

Analog-to-Digital Conversion

Chapter 1

Introduction

Even in today's highly digital world, analog-to-digital (A/D) converters continue to play a vital role. When interfacing with the real world, inputs and outputs are usually analog. To deal with this, engineers need to be skilled in both analog and digital techniques in order for a data acquisition system to truly meet performance criteria and still be cost-competitive. Competition and technology are constantly pushing the performance of A/D converters to higher levels. This performance has been enhanced by the use of digital techniques such as digital signal processing (DSP) for filtering.

Initially, we take a look at a "typical" data acquisition system and become familiar with various possible architectures. There are several methods for accomplishing A/D conversion as well as several configurations that make up the complete system. Not only are there options on how to arrange the functional blocks, but also there are options on whether each function is required. After a review of the associated tradeoffs, a decision can be made more easily as to how the overall system should look. However, an optimum solution can be achieved only by close examination of the application.

Numerous design techniques, tradeoffs, and choices are presented in this book. To assist in navigating through all these decisions, there is a checklist for setting the system requirements. This is very useful and merits a lot of thought if an optimum solution is to be achieved. Comparisons of the several types of A/D converters are also presented along with a brief description and key characteristics. The objective is to facilitate the choice of which type of A/D converter is right for a given application. Finally, the last portion of this chapter will cover the many terms and definitions associated with data acquisition (always exciting to read). This will help take the mystery out of many important A/D converter specifications.

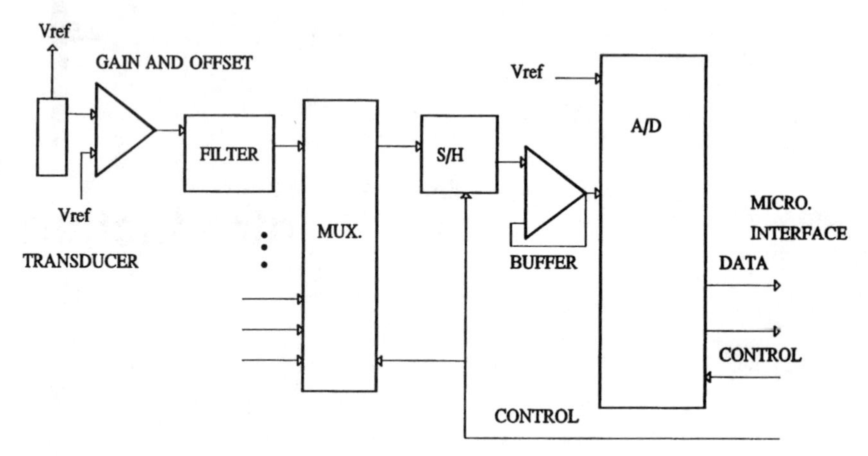

Figure 1.1 Typical data acquisition system.

Typical Data Acquisition System

When a data acquisition system is designed, there are often several options to consider. By looking at the complete picture, it should be easier to decide which basic architecture is best. Figure 1.1 illustrates a typical system. This block diagram is intended only to point out the possible blocks, and not necessarily the configuration required by a specific data acquisition application. The following sections then discuss architectural tradeoffs.

System References

The reference voltage (Vref) in the block diagram (Fig 1.1) is required for the transducer, offset stage, and the actual A/D converter to provide full-scale or 100 percent input for the A/D converter. The input transducer needs to be powered by a reference to generate a signal. If the output signal does not match the A/D input range, the signal requires some scaling. Ideally, the same reference should be used for all three areas. Otherwise, each reference change will show up directly in the total error. If all portions of the circuit use the same reference, and if the transducer changes proportionally with changes in V_{ref}, there will be automatic tracking (zero drift). For instance, if V_{ref} increased by 5 percent and this causes V_{in} to also increase by 5 percent, no error will be introduced. This type of system is said to be *ratiometric.*

One very important design criterion is the choice of whether the system can be ratiometric or must be absolute. This will ultimately have a huge impact on the final cost. Ratiometric conversions do not require tight reference tolerance, or drift, since only the ratio of two or more signal measurements is needed. Absolute measurements, however, require a stable and accurate reference for comparison to the unknown input signal. It is obvious that absolute measurements will be more costly than ratiometric ones. Absolute measurements not only may increase the cost of the reference itself (primarily for low drift), but also may require calibration for high accuracy. Calibration is likely to be required when the design needs to be accurate to more than 12 bits. Regardless of whether a ratiometric or an absolute measurement is needed, the reference-generated noise should be considered when the system calls for more than 14-bit performance or when dealing with very low-level input signals.

Gain

Adding a gain stage for the transducer can be optional, and it usually comes with some drawbacks. The main purpose of the gain stage is to achieve the maximum A/D resolution by scaling the input signal to match the full A/D input range. There is an alternative for achieving the desired resolution, however. Instead of adding either a fixed or a programmable gain amplifier, an A/D converter can be selected with more dynamic input range (resolution). Even though some resolution may be wasted, the A/D converter may prove to be higher-performance and in some cases may have a lower overall cost.

As shown in Fig. 1.2, either gain stages can be placed before each multiplexer input (usually a fixed gain for each transducer), or a single gain stage can be placed after the multiplexer [usually a programmable gain amplifier (PGA)]. If only one gain stage is used downstream of a multiplexer, care must be taken when low-level signals are measured. Otherwise, problems caused by adjacent-channel crosstalk and common mode to signal can result. Chapter 3 discusses how to deal with this problem.

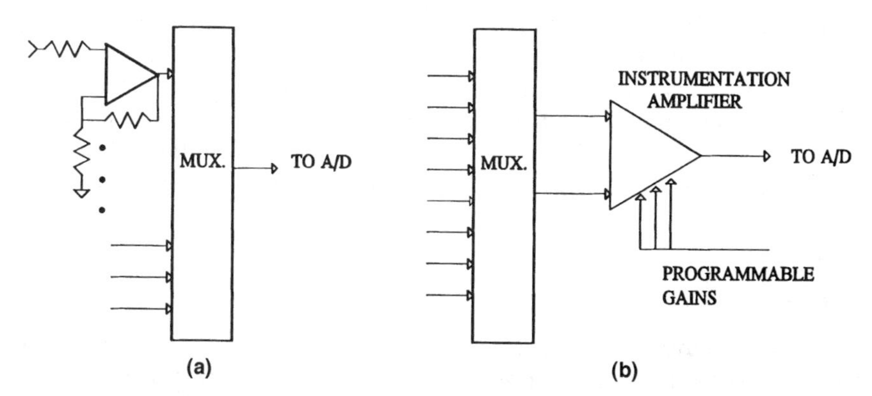

Figure 1.2 System gain options. (*a*) Fixed gain; (*b*) programmable gain.

A major drawback of using a gain stage is either the expense of tight-tolerance components (primarily resistors for setting the gain) or the required calibration due to component mismatches of limited-accuracy resistors. Even if an integrated instrumentation amplifier is used, the internal resistances have an initial accuracy that may require some form of calibration for high-accuracy designs. Still, calibration may not be an issue if the transducer or reference initial tolerance is unacceptable and must be adjusted regardless. In addition, each gain stage will produce errors from nonlinearity, noise, and drift. Among the tradeoffs are the net-cost difference between using the various gain amplifiers and the higher-resolution A/D converter as well as the level of integration (board space for each). When an instrumentation amplifier is required, a choice must be made between a lower-cost discrete or fully integrated design.

Multiplexers

Multiplexers are an option when more channels need to be measured than the A/D converter can handle. This can be a cost-effective solution, but it can introduce some problems, the most severe of which is crosstalk. Crosstalk is frequency-dependent and is the amount of feedthrough signals from unselected channels that corrupt the selected signal measurement. When low-level signals are measured, it is advisable to place gain stages prior to the multiplexer to lessen the effects of crosstalk. Very high-accuracy A/D systems may need to tolerate the added cost and use a separate A/D converter for each input, to avoid problems with multiplexer crosstalk.

When several channels need to be measured, simply tying together the outputs of the different multiplexers may not be advisable. This can cause heavy capacitive loading on the selected channel. When this occurs, the input signal will be frequency-limited from the low-pass filter created by the total capacitance and the switch-on resistance. By using a multitier architecture as shown in Fig. 1.3, the capacitance can be lowered significantly. In this example, there are 64 channels from 8 multiplexers feeding a single 8:1 multiplexer. Therefore, the maximum loading is 8 outputs per channel instead of 64. This technique is commonly used in analyzers or dedicated test equipment where multiple inputs must be measured as quickly as possible.

Filters

Filter stages are often required to attenuate the out-of-band noise to prevent aliasing or to remove system transients from the signal.

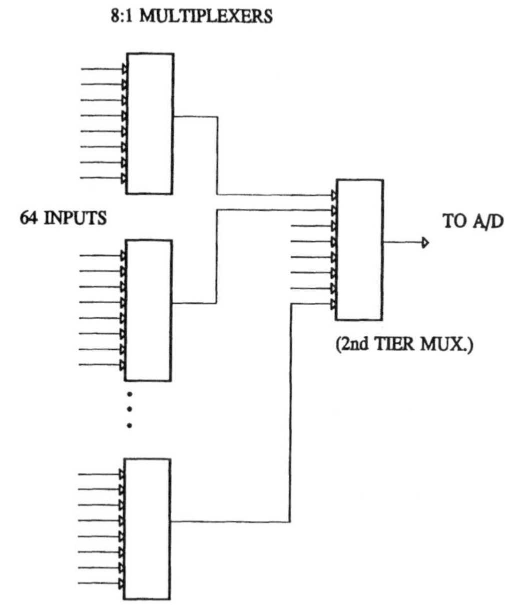

Figure 1.3 Multitier multiplexing.

Basically, aliasing is a condition in which the input signal cannot be reproduced accurately, and this will occur if the A/D sampling frequency f_s is not at least twice the maximum filter cutoff frequency f_c. Aliasing is defined in greater detail later in this chapter in the section "Understanding Data Acquisition Terms." Filter stages can be placed either before or after a multiplexer. The advantages of using a filter for each input stage include the maximum speed due to zero delay in switching channels and the option to customize the characteristics for each input signal. The disadvantage of multiple filters compared to one filter following the multiplexer (besides extra cost and board space) is that different errors will be produced by each filter. This will make autozeroing more difficult and will affect the long-term drift performance.

Sample-and-Hold (S/H)

A sample-and-hold (S/H), if not built into the A/D converter, must be added if the A/D input signal amplitude can change by more than ½ least significant bit (LSB) within the conversion time of the A/D device. Again, there are options for this function. Placing the S/H before the multiplexer allows for better synchronization of the input channels. This is due to the ability of the data acquisition system to sample each signal simultaneously, something that is often useful in

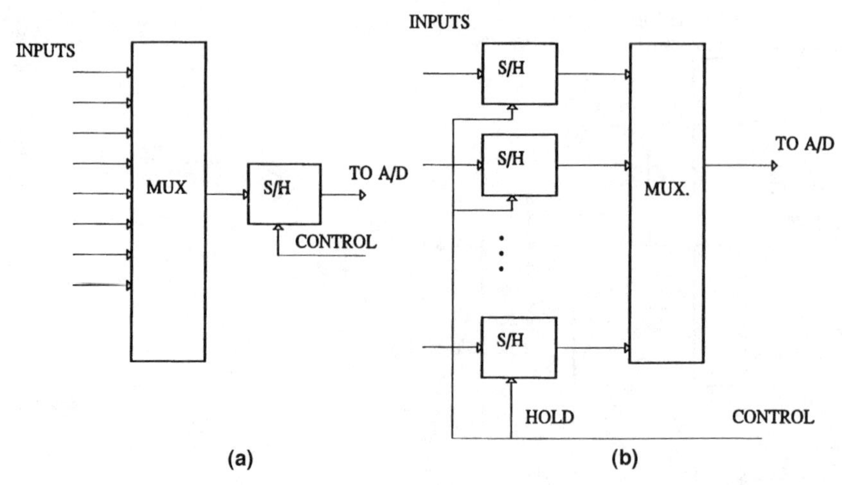

Figure 1.4 Basic S/H options. (*a*) Common S/H system; (*b*) simultaneous sampled system.

digital signal processing. An added benefit of this approach is the elimination of crosstalk errors from the multiplexer, since each input is momentarily held steady while the conversions are completed. Of course, there are also disadvantages to using several sample-and-holds. Each S/H will produce a different set of errors, with the most severe being gain variations. Figure 1.4 illustrates two options for using a S/H in the system.

A/D Converter Selection

In the previous sections we looked at numerous architectural tradeoffs leading up to the A/D converter. Choosing the best A/D converter type for a given application is far more involved and is the main focus of this book. Discussions in the following sections will provide several guidelines for making this decision. Later chapters then cover each type of converter in depth so the reader can understand how to make each converter work for a given application.

When starting a design, the engineer needs to determine the answers to several questions before attempting to select a type of A/D converter. The following checklist identifies some of the key system requirements that should be considered.

A/D converter checklist

1. *How many bits?* This refers to the amount of resolution required for a signal measurement. For example, determine the A/D

resolution required for measuring temperature from − 40 to + 85°C in 0.5°C increments. Since this requires 125 (degrees) × 2 = 250 increments, an 8-bit converter which provides 256 levels is needed. Quite often the resolution is the most important specification of the A/D converter.

2. *What accuracy?* Accuracy requirements are usually more difficult to determine than the resolution. Here the engineer must study the application and determine what is really required. For example, in an automotive throttle control application, the throttle position is largely controlled by the driver. The sensor provides position information feedback to the microcontroller for the purposes of moving the throttle to the desired position. In this application, it is the resolution that is more important (usually 10 bits), not the actual accuracy. This is due to the fact that the position sensor is usually not very linear or precise, and the driver just wants to be able to either increase or decrease the vehicle's speed by some minimum amount (resolution).

3. *What conversion rate?* Three requirements generally determine the necessary A/D conversion rate: maximum measured signal change in a given time, complete system workload (available bandwidth), and how fast the system can (or needs to) respond. In motor control applications, e.g., current measurements can be used as the controlling variable. Electric motors often have very low bandwidths (i.e., 10 Hz) and cannot respond to fast changes in drive. Unless other system requirements demand a fast conversion, the current measurement can be made within several milliseconds. On the other end of the spectrum, if the system requires DSP operations on a high-frequency input signal, then the fastest A/D converter available is potentially needed.

4. *Total system error budget?* This must include everything from the transducer up to the A/D converter. These errors include the direct-current (dc) tolerances, alternating current (noise), temperature drift over full operating range, and drift related to time. Here, decisions need to be made as to where to spend your error budget. For cost and performance reasons, keep in mind that the processor can often perform autozero operations that can significantly reduce the total error.

5. *Input signal range?* Depending on how the input signal range matches the chosen A/D full-scale input, a gain and/or offset stage may be required. Don't forget that there is an option to use a higher-resolution A/D converter rather than add gain stages—providing that 0 percent input matches the A/D converter zero level. Otherwise, an amplifier with a zero offset function will be required, and adding gain is not an issue.

6. *Total system cost target?* This usually requires a thorough analysis of the total system design so that the proper tradeoffs can be made. Besides hardware cost, there are manufacturing concerns (i.e., number of components and calibration).

7. *Input impedance?* Buffering will be required if the input signal is generated from a high-impedance circuit. This can cause problems due to the A/D converter input capacitance and its associated delays or switch currents.

8. *AC or dc (low-frequency) input?* Depending on the desired input frequency content, either the dynamic (ac) or static (dc) specifications will be more important.

Overview of Various A/D Converter Types

There are basically 12 types of A/D converters discussed in this book. Each conversion method has its own set of advantages and disadvantages that must be weighed carefully. As a start, let's briefly take a look at each type to get an idea of how it works and where it might fit.

Flash converter: Highest-speed A/D converter that uses $n - 1$ comparators (where n = bits of resolution, i.e., 8 bits = 255 comparators!) along with a decoder. The decoder simultaneously reads each comparator with the result determined by the highest-priority comparator. These types of converters are commonly used in automatic test equipment where the high cost can be justified.

Subranging converter: Very similar to flash converters, except that these perform two or more substeps or multisteps instead of requiring a complete conversion in one cycle. Although not as fast, the subranging converter is much less expensive than a full flash converter because significantly fewer comparators are required.

Delta-sigma converter: Provides very high resolution by oversampling and modulation (feedback integration). It utilizes low-performance analog circuitry by shifting the burden to the digital circuitry with the aid of very large-scale integration (VLSI). This method has become an attractive choice for moderate-cost dedicated applications such as instrumentation and audio.

Dual-slope converter: This extremely high-resolution method works by performing an integration of an unknown input signal, followed by the integration of a known reference signal of opposite polarity. Time measurement comparisons of the two ramp signals are then used to determine the unknown input signal level, the time for which can be very long (several seconds for high resolution, i.e., 20 bits). This technique is very commonly used in all forms of instrumentation (i.e., including weighing scales).

Single-slope converter: Comparable to the dual-slope, but operates with only a single, fixed slope for a complete measurement. The time for the fixed ramp to equal the input signal(s) at a comparator is directly proportional to the magnitude(s). Infrequent measurements of a reference signal provide an absolute measurement, while ratiometric measurements do not require any reference measurements. This results in a slightly faster conversion (between 2 and 3 times faster). Although not quite as accurate as the dual-slope converter (due to system noise integration), the single-slope converter is much less expensive.

Sampling (Successive-Approximation) converter: This is by far the most popular fully integrated solution for moderate-performance systems. This works by approximating the input with a digital-to-analog converter (DAC) and testing each bit, starting with the most significant bit (MSB) and ending with the least significant bit (LSB).

Voltage-to-frequency (V/F) converter: Changes in input voltage, resistance, or capacitance are transformed to changes in output frequency. Since only one input (usually to a timer) is required, the signal can be easily measured by the microcontroller. The V/F converter achieves great resolution, but suffers from several errors (mostly with the circuit capacitance).

R-2R converter: By utilizing either thick- or thin-film *R*-2*R* resistor networks to create a DAC, a very low-cost successive-approximation A/D converter can be created. With a little help from the microcontroller (low software overhead) in driving the *R*-2*R* ladder network, a reasonably fast converter can be formed.

RC converter: This is a nonlinear version of the single-slope converter. However, with the aid of some software techniques, this can be made rather easy to deal with. Largely as a result of the capacitor, the accuracy is limited.

Resistance measurement converter: Although this is not an A/D converter in the strict sense, this technique can very easily and accurately determine the percentage of an input potentiometer. The technique either charges or discharges a capacitor with a known resistance and then an unknown resistance. The linear result is the ratio of the two times.

Pulse-width modulation (PWM) converter: By using a fixed period and varying the duty cycle to a low-pass filter, the average voltage created by the duty cycle is compared to the unknown input voltage. This method is practical only for the slowest systems of low to moderate resolution.

Improved PWM converter: Instead of using a single duty cycle of high resolution, several shorter pulses are applied to a low-pass filter. This method is significantly faster than the conventional technique, and it more closely resembles the delta-sigma method.

Comparison of Key A/D Converter Characteristics

In making a comparison of the available types of A/D converter architectures, usually some key requirements can be used to quickly sort out optimum approaches. The most important characteristics include resolution, accuracy, speed, and cost. The following comparisons are a good estimate of these key requirements. Keep in mind that these criteria may need to be expanded based on the actual design. In some cases the accuracy comparisons assume that either an autozero or a manual calibration is performed when required. Additionally, conversion times will often depend on the frequency of operation (oscillator clock) and the desired resolution.

The first comparison is of cost versus accuracy, and this is listed below, starting with the most costly to the least expensive:

Cost	Maximum accuracy, bits	Type of A/D converter
Highest	16	Flash
	18	Delta-sigma (oversampling)
	16	Subranging (multistep)
	22	Dual-slope (integrating)
	16	Sampling (successive-approximation)
	14	Single-slope
	8	Voltage/frequency (calibration required)
	8	*R*-2*R* (successive-approximation)
	4	*RC* (resistor capacitor time constant with 1 percent V_{ref})
	10	PWM (pulse-width modulation)
	10	Improved PWM (same cost as standard PWM)
Lowest	7	Resistance measurement

The next comparison is of speed versus resolution. Note, in most cases, that the speed is proportional to the timer oscillator frequency and the desired resolution.

Speed, conversions per second	Resolution range, bits	Type of A/D converter
10^6–500×10^6	6–16	Flash
10^4–10^5	12–22	Delta-sigma (oversampling)
150,000–10^7	8–16	Subranging (multistep)
10–30	12–24	Dual-slope (integrating)
10^4–10^6	8–16	Sampling (successive-approximation)
60–40,000	8–14	Single-slope
60–40,000	8–14	Voltage/frequency (calibration required)
10,000–40,000	6–10	*R*-2*R*
4000–30,000	4–8	*RC* (resistor capacitor time constant)
2–10	6–10	PWM
50–4000	6–11	Improved PWM (same cost as standard PWM)
4–15,000	6–10	Resistance measurement

Understanding Data Acquisition Terms

The following terms are organized in alphabetical order in two sections. First the definitions for A/D converters in general are presented. Then the terms normally associated with sample-and-holds are explained.

A/D terms

Aliasing. Based on the sampling theorem, an input signal must be sampled at a rate at least twice its maximum frequency component. Otherwise, distortion will result, and the signal measured cannot be completely recovered. When the A/D system samples the input at a rate f_s, the original spectrum is shifted out at multiples of $n \times f_s$ (where $n = 1, 2, 3, \ldots$). As shown in Fig. 1.5, if the input signal is not

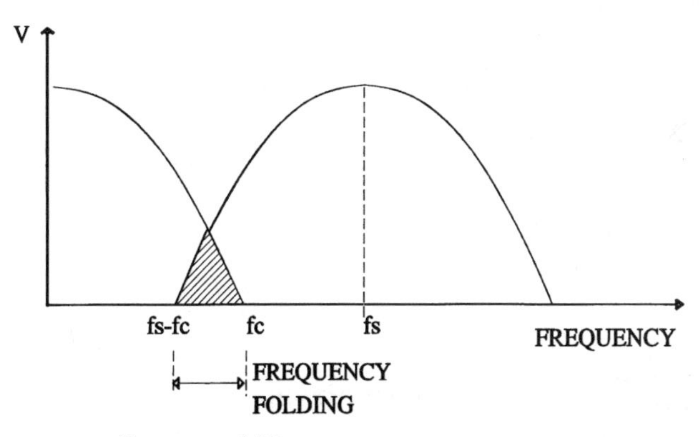

Figure 1.5 Frequency folding.

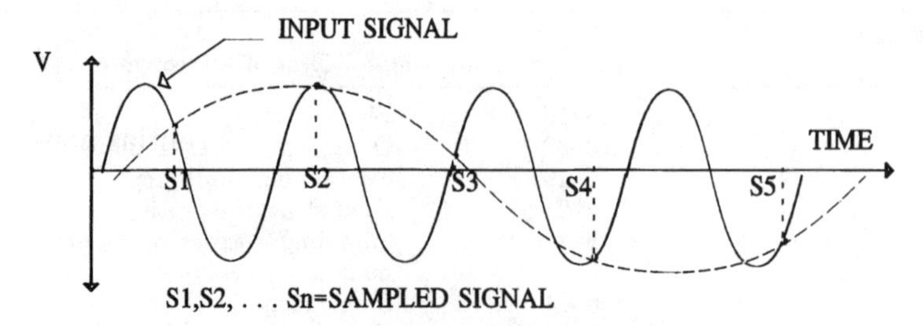

Figure 1.6 Aliasing within the frequency spectrum.

bandwidth-limited to below f_s /2, there will be a crossover (also known as *folding*) between the sampled signal and the original signal. In practical applications, the input signal must be sampled at a rate much faster than f_c to avoid problems with system noise and less-than-ideal filter responses. Figure 1.6 illustrates the problem caused by inadequate sampling of the input signal. Note that the measured signal appears to be much lower in frequency than the original signal.

Differential nonlinearity (DN). When consecutive conversions change state from an adjacent level by 1 LSB, there is no differential error present. On the contrary, any deviation from one adjacent level by more or less than 1 LSB is the amount of differential linearity error. For example (see Fig. 1.7), if a level required $1\frac{1}{4}$ LSB more than the

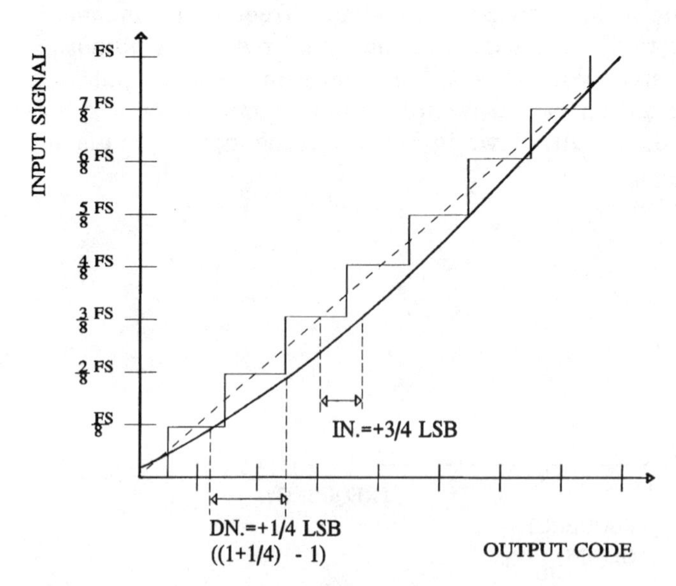

Figure 1.7 Integral and differential nonlinearity.

previous level, there would be ¼-LSB differential error. When the DN reaches 1 LSB or greater, there will be missing output codes.

Dynamic differential nonlinearity. This is the same as differential nonlinearity but with an alternating-current (ac) input (usually near the maximum sampling rate) instead of a direct-current (dc) input.

Dynamic integral nonlinearity. This is the same as nonlinearity but with an ac input instead of a dc one.

Dynamic specifications. These refer to the ac performance specifications (as opposed to traditional specifications with dc inputs). Included are *S/N,* SINAD, ENOB (which includes dynamic differential and integral nonlinearity), THD, FPBW, and SSBW.

Effective number of bits (ENOB). The ENOB, expressed in decibels, takes into account several dynamic specifications and is determined by the resolution of the A/D converter.

$$\mathrm{ENOB} = \frac{\mathrm{SINAD} - 1.76}{6.02}$$

Full-power bandwidth. This is defined as the frequency at which the input signal near the A/D full-scale value has caused the signal-to-noise ratio (SNR) to decrease by 3 dB (ENOB has dropped by ½ bit).

Gain error. This is also called the *full-scale error,* and it refers to the difference between the input that produces a full-scale code and the ideal voltage, expressed in least significant bits (Fig. 1.8).

Integral nonlinearity. This is the worst-case deviation from a straight line between the zero and full-scale endpoints, expressed in LSBs and shown in Fig. 1.7.

Intermodulation distortion (IMD). When two different-frequency signals are present, there will be an interaction caused by the A/D nonlinearity that generates additional frequency components. These additional frequencies will consist of the sum and difference between the original input frequencies and their harmonics.

LSB (least significant bit). One LSB = (full-scale input voltage)/resolution. For example, with V full scale = 5 V and resolution = 10 bits, then 1 LSB = 5/1024 = 4.9 mV.

MSB (most significant bit). The highest weighting factor of the output code equals ½ full scale.

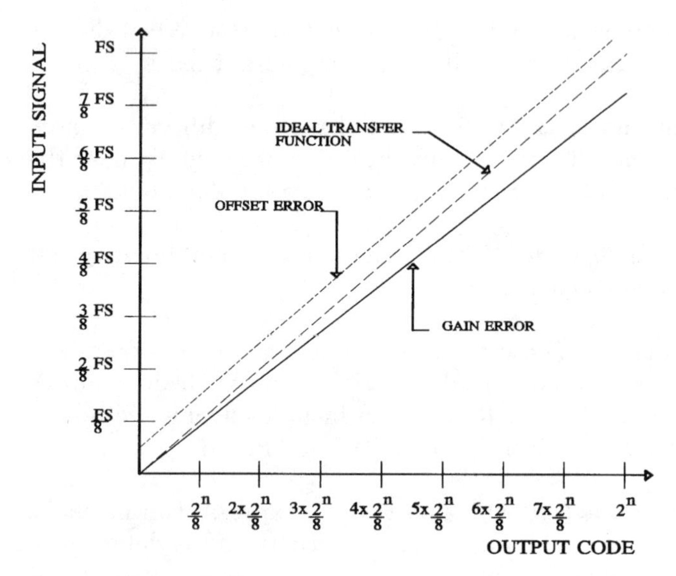

Figure 1.8 Gain and offset errors.

Offset error (zero error). As shown in Fig. 1.8, this is the shift in the line drawn between the midpoints of the converter thresholds to the ideal response in LSBs.

Peak harmonic. This is the ratio of the magnitudes of the next largest harmonic (usually the second) to the fundamental.

Peak harmonic or spurious noise. This is the same as above, but compared to noninteger multiples of the fundamental. This is usually the result of harmonics of aliased high-input-frequency inputs.

Power supply sensitivity. This is the change in output codes for changes in supply voltage.

Quantization error. All A/D converters will have at least a minimum error as a result of the discrete steps that represent the analog input, and this error is directly proportional to the resolution.

$$\text{Quantization error} = \pm \tfrac{1}{2}\ \text{LSB}$$

Ratiometric measurement. This is a measurement between two or more signals based on only the ratio of the signals to each other, and not compared to a fixed reference signal. In practical terms, this means that (1) the input signal is powered from the same reference signal as the A/D converter and will change proportionally or (2) all signals are measured with respect to the same reference signal, and the ratios are computed accordingly.

Resolution. The number of bits of resolution refers to the smallest input level to cause change in the output code. For example, a 10-bit converter has 2^{10} or 1024 unique levels.

Sampling rate. This is the inverse of throughput time and is a measure of the maximum input signal frequency that can be accurately measured.

Signal-to-noise ratio (S/N). This is the ratio of the input signal S to the background noise N in a system. For an ideal A/D converter with a sine wave input, the SNR related to the resolution n is

$$\mathrm{SNR}_{\mathrm{rms}} = 6.02n + 1.76 \qquad \mathrm{dB}$$

SINAD (ratio of signal to noise plus distortion). This includes the SNR along with the total harmonic distortion (THD) of the A/D converter.

Small-signal bandwidth. This is rather loosely defined as the frequency at which the SNR has decreased by 3 dB with an input signal much less than full-scale.

Spurious free dynamic range. This is the ratio of the full-scale input signal to the highest harmonic or spurious noise component amplitude. Essentially, this is an indication of how far it is possible to go below the full-scale input signal without hitting noise or distortion.

Throughput time. This is the total time it takes for an A/D converter to complete consecutive measurements and generate the output code.

Total harmonic distortion (THD). A very important specification in audio systems, the THD is defined to be the rms ratio of the sum of the harmonic distortion amplitudes to the original input amplitude. It is caused by the A/D converter nonlinearities.

Total unadjusted error (TUE). This is more or less a catchall term for the worst-case dc errors which include offset, gain, and differential and/or integral nonlinearity errors.

S/H terms

Acquisition time. This is the total time for the sample-and-hold to acquire (or charge the capacitor) a full-scale step input signal on the

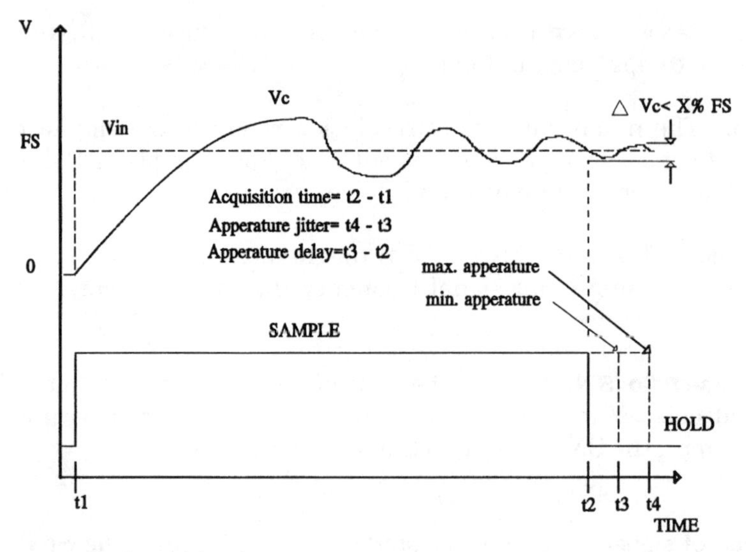

Figure 1.9 Acquisition time.

holding capacitor within some specified percentage of full scale or LSBs. This is for the transitions from t_1 to t_2, as shown in Fig. 1.9.

Aperture delay time. This is the delay from the time that the hold command is applied to the time when the switch actually opens.

Aperture jitter or uncertainty. This is the error due to circuit noise and, to a lesser extent, jitter in the external clock signal (a system problem). All sample-and-holds will have some variation (jitter) between samples as to when the signal is actually sampled upon application of the hold command. Figure 1.9 illustrates this very important specification in measuring high-frequency input signals.

Aperture time. This is the time for the sample-to-hold transition. The amount of error produced will be proportional to the frequency and amplitude of the input signal.

Droop rate. Hold capacitors will droop or lose their charge over time as a result of circuit leakage current. When the internal capacitance of the S/H is used, the data sheet specifications will describe the expected drop. Using an external capacitance will improve the droop rate as described below, based on the basic equation of $I = C\,dV/dt$.

$$V_{\text{droop}} = \frac{(I_{\text{leakage}})(\text{time})}{C_{\text{hold}}}$$

Feedthrough error V_{ft}. When in the hold mode, this is the amount of input signal that works its way to the output through the source-drain capacitance (C_{ds}) of the hold switch and is proportional to the hold capacitor C_h and dV_{in}:

$$V_{ft} = \frac{C_{ds}\, dV_{in}}{C_h}$$

Gain error. Since all sample-and-holds have gain stages, the resulting dc gain error is defined as the ratio of input to output signal range. This can be a significant portion of the total system dc error.

Hold-mode droop. Once the input signal is held on the capacitor, there will be some droop (or loss of signal) as a result of the circuit leakage current.

Hold-mode feedthrough. This is defined as the percent of input signal from a step change or frequency that shows up in the output while the switch is open (hold mode).

Hold-mode settling time. This is the time from the hold command to the time when the output signal settles to within some specified percentage (see Fig. 1.9) of a near-full-scale input step.

Sample-to-hold offset error V_{hs}. This is also known as the *pedestal offset* and is the offset caused by switched currents (transfer of charge) into the holding capacitor (C_h) during hold transition (opening of the switch). The magnitude of V_{hs} can be predicted by

$$V_{hs} = \frac{Q}{C_h}$$

Sample-to-hold transient. This is a small spike that appears at the output, and it is caused by feedthrough of the sample-to-hold logic signal.

Chapter

2

Passive Support Components

More than likely, precision integrated circuits will end up depending on the passive components that surround them for accuracy as well as stability. For this reason, an engineer is wise to fully understand the limitations associated with the less glamorous components such as resistors and capacitors. It is easy to overlook these seemingly simple devices because of their basic properties. However, as will be pointed out in this chapter, there are a wide variety of characteristics and performance ranges for resistors and capacitors. If these attributes are not understood, the circuit performance can (and probably will) be severely affected. After a review of the important device characteristics, it will be easier to make an intelligent choice of the right part for the application.

When an amplifier circuit is designed, the closed-loop gain eventually is limited to the accuracy of the network resistors used. Besides the initial tolerances and ratio matching, the resistor stabilities are very important. Some of the not-so-obvious causes of resistance changes come from high-temperature soldering, humidity (moisture absorption), and the voltage coefficient. In addition to the resistance tolerances, cost is usually a major factor in the design of a precision circuit. Continued advances in thin-film technology have produced both discrete and network resistors that are highly accurate in most respects and provide a low-cost alternative to wirewound resistors.

Capacitors are available in a variety of materials with diverse characteristics for specific applications. For instance, the most common use of capacitors is for power supply filtering. In these applications it is necessary to both filter the incoming power and provide the instantaneous load current. Thus the series inductance and resistance are important so that current can be supplied without large supply voltage drops. Other uses rely on stability and accuracy for functions such as timing, integrators, and active filters. Given that each application has its own set of requirements, capacitors must be chosen carefully. Capacitors are not nearly as accurate—or available in as many values—as resistors. For this reason, circuits requiring capacitors for precision functions generally are more difficult to design and usually require some form of calibration.

Resistive Components

Resistors are available in different forms (discrete, network, potentiometer, and as part of an integrated circuit) and have a very wide range of values, initial tolerances, drift, and wattage. In addition, there are several choices to make from the list below of available materials used for creating resistors.

- Discrete resistors
  - Carbon composition
  - Carbon film
  - Metal film
  - Thin film
  - Wirewound
  - Foil
- Networks
  - Thick film
  - Thin film
- Potentiometers
  - Cermet
  - Conductive plastic
- Wirewound

Before we describe each type of resistor, we will explain the important parameters. Some of the terms listed here will not be found on every manufacturer's data sheets. Unfortunately, there is a lack of standardization in specifying resistors. However, most of the significant specifications are usually stated or can be obtained from the manufacturer.

Resistor specifications

1. Accuracy = maximum difference from the nominal resistance value, stated in percent.

2. Temperature coefficient = maximum change in resistance over a specified temperature range, stated in parts per million (ppm).
3. Time drift = amount of drift from initial value to final resistance value, usually stated in 1000 hours or 10,000 hours of continuous operation.
4. Voltage coefficient = maximum change in resistance due to applied voltage, stated in parts per million per volt. This does not include the effects of self-heating. Changes are due to molecular distortions from applied voltage.
5. Thermal noise voltage = self-generated noise, expressed in rms or microvolts peak to peak. This is discussed in detail in Chap. 9.
6. Self-heating = change of resistance due to power dissipation, specified in parts per million per watt. Using the thermal resistance (degrees Celsius per watt) and the temperature coefficient will produce the same result.

Example: Assume an 80°C/W thermal resistance, a temperature coefficient of 25 ppm/°C, and a power dissipation of 0.25 W. Determine the change in resistance.

$$\text{Resistance change} = 100[(80°\text{C/W})\,(25\ \text{ppm/°C})\,(0.25\ \text{W})]$$
$$= 0.05\%$$

7. Thermal shock or resistance to soldering heat (e.g., 350°C) is the maximum change in resistance after soldering to circuit board.

The various resistor attributes and construction methods are described below as an aid in determining the best choice for a given application.

Discrete resistors and networks

Carbon composition. Carbon-composition resistors are made of a solid resistive core and exhibit relatively poor noise, poor temperature coefficient, and high breakdown voltage. They are the lowest-cost alternative.

Carbon film. This type is made by depositing resistive material on a nonconductive core, such as ceramic or glass, and cutting a spiral pattern on it to achieve the desired resistance. Due to the small gaps between the cuts, the voltage breakdown is much less than with the carbon-composition resistors. Although available in tolerances down to 1 percent, these types have high temperature coefficients that can cause unwanted drift errors. Carbon films have relatively low to medium cost.

Metal film. Metal film is similar in construction to carbon film, but with a spiral-cut metal film. Metal-film resistors offer significantly better temperature coefficients and slightly better noise than carbon-film ones. Metal-film resistors offer an economical solution for precision circuits.

Thin film. These resistors are used for both discrete and networks. They are created by depositing a thin film (10 to 100 nm) of nickel-chromium or tantalum nitrate on a flat substrate. Matching the thermal characteristics of the substrate to the film is vital to achieving low temperature coefficients. Thin-film resistors are preferred for high-precision applications because tightly controlled specifications provide high initial tolerance and close tracking. Recent advances have produced performances equaling all but the most accurate wirewound resistors. Cost is medium to medium-high for very high precision.

Thick film. This is also used for discrete and networks. It is manufactured by applying a relatively thick (about 10 μm) paste or ink consisting of metal/metal oxide mixed with glass and resin for binding. Although not as accurate as thin film, thick film provides a wider range of resistances at a lower cost. When used for networks, thick films provide good ratio and temperature-tracking performance.

Wirewound. These resistors are produced by winding several loops of wire (nickel-chromium) on a nonconductive core. Although very accurate with the lowest noise, these resistors have the drawbacks of high cost, larger packages, and much higher series inductance. Another drawback is the higher distributed capacitance from the large number of windings, which can exceed 1000 turns.

Foil. Foil resistors are formed by bonding a nickel-chromium foil to a ceramic substrate and etching a winding pattern to achieve a desired resistance. Their attributes include highest stability and medium cost. In comparison to wirewound resistors, foil resistors provide much less inductance, but generally not as high resistance values.

Potentiometers

Cermet. Ordinarily this is a single-turn potentiometer made with a thick conductive film. Cermets provide nearly infinite adjustability with good linearity, and they are available in a wide range of values. These have the advantage of better temperature drift and higher moisture resistance.

Conductive plastic. This potentiometer is similar to cermet in construction, but differs in having superior rotational smoothness for longer life than cermet or wirewound potentiometers.

Wirewound. As the name implies, this type is made by winding several turns of wire around a nonconductive material. The advantages of wirewound potentiometers include good stability, low noise, and excellent linearity. The disadvantages are due to the large number of turns. This causes small jumps in resistances as the potentiometer is adjusted or is subjected to vibration. Higher inductance and higher capacitance are another result of the large number of turns which can limit the frequency range (especially with high-value potentiometers).

Rather than list the important specifications separately and make a comparison, consulting the attributes listed in Table 2.1 should make the job much easier. Keep in mind that these are intended to provide a guide for proper selection, and that each manufacturer may have slightly different performance levels for a similar device.

Resistor applications

Resistors are primarily used in precision circuits for setting the gain of an amplifier or for setting a reference level by a resistor divider. There are a few things that should be considered when resistors are applied. Besides the more obvious initial tolerance of the resistors, the stability of the resistor is usually far more important to consider. This is where resistor networks have a significant advantage over discrete resistors. Due to the common substrate and manufacturing process, networks track extremely well. For instance, thin-film networks from Electro-Films Inc. are available with tracking temperature coefficients as low as 0.5 ppm/°C. If standard networks are not suitable, most manufacturers will make custom networks (configurations and values) for a very low engineering cost. Thick-film networks are also attractive even though their absolute accuracy is no better than 1 to 2 percent. This is because of the ratio matching being much closer, which is usually all that is important. Some common applications for networks in precision circuits are shown in Fig 2.1. In these applications, the ratio matching and tracking are critical for performance.

Potentiometers are commonly used for making either slight zero or gain adjustments for precision analog circuits. In designing with a potentiometer, be aware that the initial and temperature tolerances are typically not very good (that is, ±10 percent and 150 ppm, respectively). There are also several effects from vibration to account for.

TABLE 2.1 Comparison of Typical Resistor Characteristics

Type	Tolerance range, %	Ratio tolerance, %	Absolute TC, ppm/°C	Tracking TC, ppm/°C	Stability, %/1000 h	Resistance range
Carbon film	5	N/A	200–500	N/A	2	10–50 MΩ
Metal film	0.1–1	N/A	5–100	N/A	0.1	1–10 MΩ
Carbon composition	2–30	N/A	to 1000	N/A	—	0.1–0.1 GΩ
Wirewound	0.01–0.1	N/A	1–10	N/A	0.002	0.01– MΩ
Thick film	0.5–2	0.5–2	100–300	5–50	—	1–10 MΩ
Thin film (NiCr)	0.05–0.5	0.003	10–50	0.5	(Ratio-Abs.) 0.005–0.02	1–1 MΩ
Thin film (TaN)	<0.1–1	0.02	10–100	2	(Ratio-Abs.) 0.005–0.02	1–25 MΩ
Foil	0.025–0.1	N/A	0.5–2	0.5–2	0.0005	to 250 kΩ
Cermet potentiometer	3–20	N/A	100–200	N/A	3	10 Ω–1 MΩ
Conductive plastic potentiometer	5–20	N/A	1000	N/A	10	to 2.5 MΩ

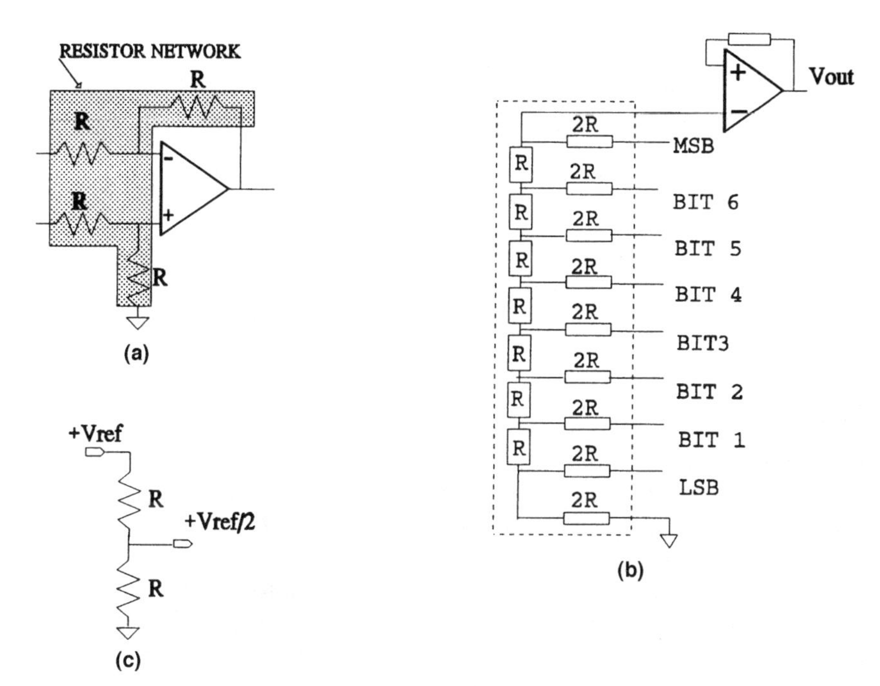

Figure 2.1 Common resistor network applications. (*a*) Discrete differential amplifier; (*b*) R-$2R$ discrete DAC; (*c*) reference-level generation.

Because of this, it is recommended that potentiometers be used only if absolutely necessary. If they are required, always limit the potentiometer to provide only the adjustment required. For example, in Fig. 2.2 there is a conductive plastic potentiometer used for fine-trimming an amplifier gain. Here, it is only required to take up the 1 percent tolerances of the discrete resistors. With the potentiometer sized to provide a maximum adjustability of about 5 percent, the net result of high-temperature drift (that is, ± 1000 ppm) will be reduced to less than 50 ppm.

Another important consideration in using a potentiometer is the need to limit the wiper current. To eliminate the possibility of damage, never place a constant voltage between the wiper and one of its endpoints. If the wiper is adjusted down to an endpoint, a large current can damage the potentiometer. Placing a resistor in series with the wiper will solve that problem. Depending on the type of potentiometer used and the number of adjustments expected, it may be a good idea to also use a resistor in parallel between the wiper and an endpoint. This will ensure the desired operation in the event of a wiper open circuit.

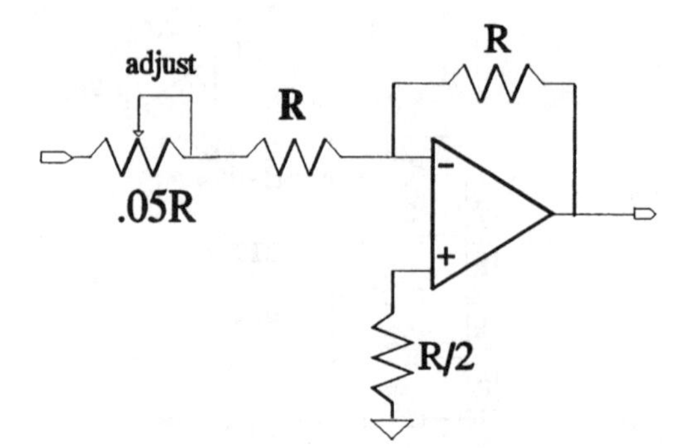

Figure 2.2 Potentiometer amplifier gain adjustment.

Capacitors

Not only are there many different capacitor types, but also there are broad ranges in characteristics within a given family, or among manufacturers. This can make the selection of an optimum capacitor for an application quite difficult. Circuits that depend on the accuracy of capacitors are much more difficult to design due to the limited values and less accurate characteristics compared to resistors. The following list identifies typical applications for capacitors in precision circuits and the various capacitor families.

1. Capacitor applications

   a. Timing (i.e., integrator, *RC* oscillator)
   b. Decoupling power supply ripple
   c. Supplying instantaneous integrated-circuit (IC) load current
   d. Sample-and-hold

2. Capacitor families

   a. Ceramic
   b. Film
   c. Electrolytic
   d. Tantalum

Within the above list of capacitor families there are several types with vastly different characteristics. This is especially true for ceramic and for film capacitors primarily resulting from the various dielectrics. Equivalent capacitor models shown in Fig. 2.3*a* and *b* illustrate the true circuit elements to watch out for in making a capacitor selection. This, combined with the following list of important capacitor specifications, should make the selection process easier.

Important capacitor specifications include the following:

1. *Capacitance range.* The dielectric constant for a given capacitor determines the practical size and range available.

DEVRY INSTITUTE OF TECHNOLOGY LIBRARY MISSISSAUGA, ONTARIO

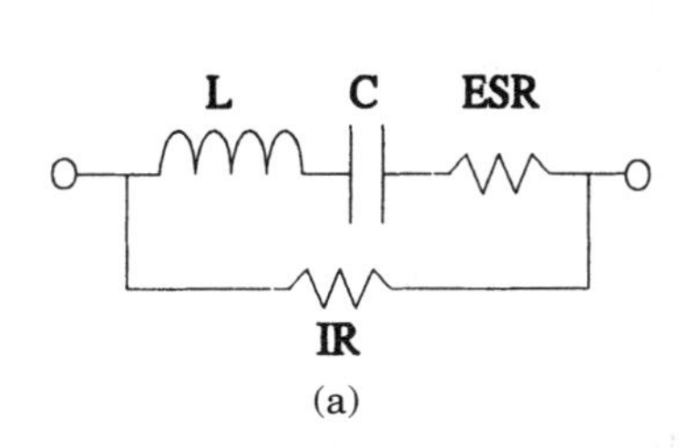

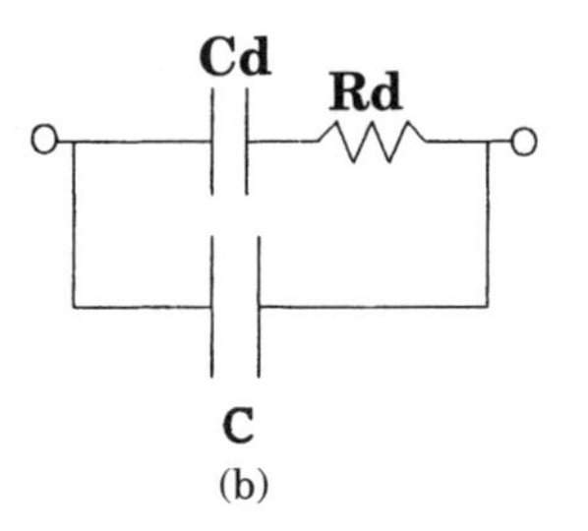

Figure 2.3 Equivalent capacitor models. (*a*) Basic circuit; (*b*) dielectric absorption. *Note:* C = nominal capacitance; L = total series inductance (i.e., leads and conductors); ESR = equivalent series resistance; IR = insulation resistance; C_d, R_d form the dielectric absorption constant.

2. *Temperature range.* The operational temperature is limited due to certain parameter degradations.
3. *Temperature drift.* This is stated in parts per million or percent over the operational temperature range.
4. *Insulation resistance (IR).* This parallel leakage resistance is usually rated by the product of ohms and capacitance. For example, a 500-MΩ · μF rating for a 0.1-μF capacitor yields a leakage resistance of 5000 MΩ.
5. *Dissipation factor (DF).* This is a measure of the losses due to the capacitor equivalent series resistance (ESR), which is a function of frequency (i.e., 120 Hz):

$$\text{ESR} = \frac{\text{DF}}{2 \times 3.14 \times f_c}$$

6. *ESR.* This is the total series resistance of the leads and conducting electrodes (frequency-dependent).
7. *Dielectric absorption (DA).* This is the amount of charge absorbed within the dielectric material that reappears on the conductive plates following a change in capacitor voltage. It is expressed in percent.

The following sections briefly describe important characteristics of available capacitors along with their typical applications, starting with ceramic, film, aluminum electrolytic, and finally tantalum capacitors.

1. *Ceramic capacitors.* These are available in solid (disk) or multilayer construction, providing low inductance and ESR which make them ideal for power supply filters and decoupling. They come in basically three grades of dielectric constant with varying temperature performances.

NPO/CGO = temperature-stable, good for timing applications, etc., with the lowest dielectric constant among ceramic capacitors

X7R = medium dielectric constant providing good volume efficiency and moderate temperature coefficient

Z5U/Y5V = highest dielectric constant for realization of smallest packages but has the worst temperature performance

2. *Film capacitors.* Describing film capacitors is considerably more difficult than describing other types. For one reason, there are many more types of dielectric materials to choose from. In addition, there are two different ways to produce the electrodes. Metallized film capacitors are created by depositing a metal alloy (i.e., zinc or aluminum) on a plastic film. This method provides a thin layer for higher volumetric efficiency compared to foil types. The other method of creating an electrode is to use a foil. Foil capacitors use alternately layered sheets of foil and dielectric film and are more rugged when it comes to handling high current. The common film dielectrics are listed below.

   a. Polypropylene film possesses the best characteristics when overall requirements are considered (i.e., stability, dielectric absorption, and cost). The only potential disadvantage is the lower operating temperature range (105°C).
   b. Polyester films are primarily used for direct-current (dc) applications such as filtering. These are the lowest cost of all films and provide the smallest packages for equivalent capacitance. However, polyester films have the disadvantage of having the highest temperature sensitivity and loss factor at high temperature.
   c. Polycarbonate film is better than polypropylene as far as stability and the dissipation factor are concerned and has performance approaching that of polystyrene. Another advantage of polycarbonate film is its higher upper-temperature limit.
   d. Polystyrene films have the best performance including dielectric absorption, temperature drift, and insulation resistance. For many wide-temperature-range applications, major drawbacks include an upper temperature range limited to about 85°C and higher cost.

The curves in Fig. 2.4 show some important characteristics of film capacitors as a function of temperature. It is quite often the temperature effects that create the most problems.

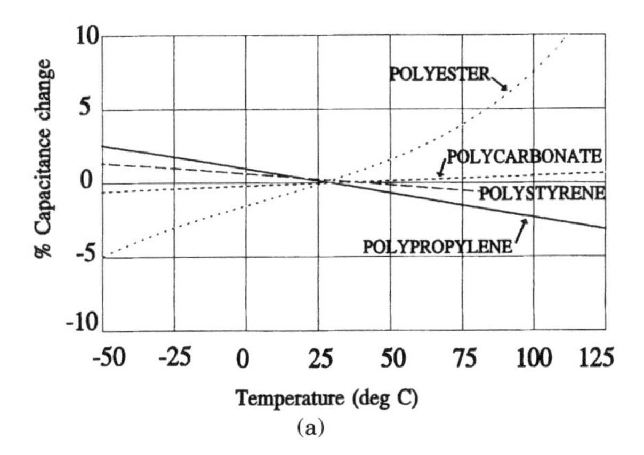

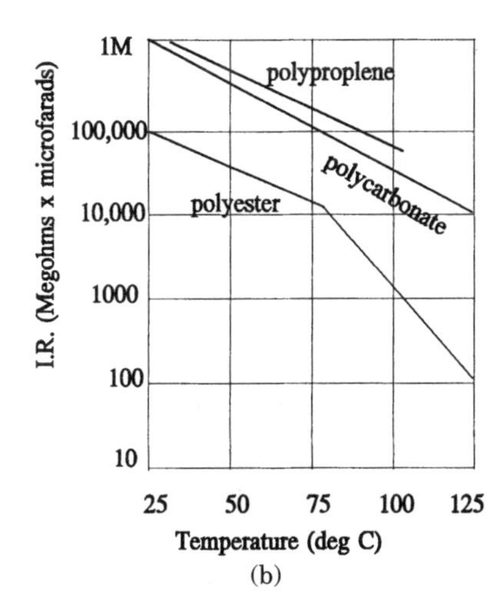

Figure 2.4 Typical film parameters versus temperature. (*a*) Percentage of capacitance change; (*b*) insulation resistance.

3. *Aluminum electrolytic capacitors.* These capacitors offer the most capacitance per cost (see Fig. 2.5). It is for this reason that aluminum electrolytic capacitors are by far the most widely used for applications needing large values (that is, >10 μF). As the name implies, these capacitors are created with aluminum electrodes which are highly etched to provide maximum surface area (more capacitance). The aluminum oxide dielectric constant of approximately 8.4 is about one-third of the dielectric for tantalum (about 26). Therefore, electrolytic capacitors have higher leakage current and higher ESR (equivalent series resistance) than tantalum capacitors. Another drawback to using electrolytic capacitors is the

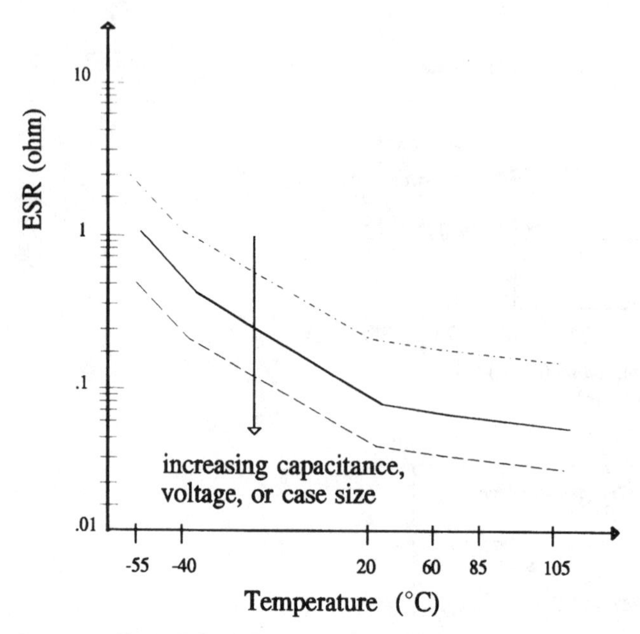

Figure 2.5 Typical electrolytic capacitor ESR versus temperature.

much higher ESR that increases significantly with lower temperatures (i.e., by a factor of 4 to 7 from +25 to −40°C). One trick to get around the high ESR is to choose a much higher capacitor than required.

4. *Tantalum capacitors.* There are three types of tantalum capacitors: solid, wet, and foil. In each case, the dielectric used is tantalum pentoxide. However, solid tantalum capacitors use a solid electrolyte of manganese dioxide that provides the lowest-cost capacitors of the three types as well as excellent low-temperature performance (ESR and stability). Wet types are primarily used for applications where the lowest possible leakage current is desired. Foil types offer advantages with their ability to operate at very high voltages (that is, 300 V) or when a substantial reverse voltage is possible.

Summarizing the capacitor characteristics will make the comparisons for proper selection much easier (see Table 2.2). It is even more important than in the case for resistors to keep in mind that this represents guidelines of typical characteristics, since even the same type can have dramatically different performances.

TABLE 2.2 Comparison of Typical Capacitor Characteristics

	Capacitor range	Temperature coefficient, %	Stability, 1000 h	Dielectric absorption, %	Dielectric constant	Dissipation factor, %	Insulation resistance	Relative cost	Temperature range, °C
Ceramic CGO	0.015 μF – 1 pF	0.5 (−40 to + 85)	0.5%	—	10	Q = 4000	10^8 MΩ	High	−55–150
X7R	150 F – 0.47 pF	±15 (−40 to + 85)	−5%	2.5	—	1	10^6 MΩ	Medium	−55–150
Z5U	56,000 F – 5.6 pF	+20/− 50 (−40 to + 85)	−20%	2.5	400,000	2	10^5 MΩ	Low	−55–150
Film									
Polyester	0.0001F – 15 μF	+10/− 3	Medium-high	0.3	3.1	0.4	25 kΩ	Low	−55–125
Polypropylene	0.001 F – 30 μF	±2	Medium	0.1	2.4	0.1	200 kΩ	Medium	−55–105
Polycarbonate	0.001 F – 22 μF	±1	Medium	0.1	2.8	0.25	100 kΩ	Medium	−55–125
Polystyrene	—	±1	Very low	0.02	2.3	0.1	200 kΩ	High	−55–85
Tantalum	0.005 F – 1000 μF	± 15	10%	N/A	26	Low	Medium	High	−55–125
Aluminum electrolytic	0.47 F – 10^6 μF	± 115	10%	N/A	8.4	High	Medium-high	Low	−40–105

Chapter

3

Active Support Components

Peripheral support components can make or break any analog-to-digital (A/D) converter design if they are not chosen carefully. In this chapter, we examine several active devices required by the A/D converter to handle various input signals. When selecting support devices, one needs to determine which dc or ac specifications are more important. These performance requirements will have a large impact on the type of reference, amplifier, comparator, sample-and-hold (S/H), and multiplexer to be selected. In situations where there is no internal sample-and-hold (S/H) provided by the A/D converter, an external S/H will likely be required. Design of this function is a difficult task and requires significant effort so that the performance of the system (both dc and ac) is not severely limited.

References

References are essential for the operation of every A/D converter to establish the full-scale input range (100 percent input scale). Additionally, the reference may be used to offset the input signal to match the A/D input voltage range, or to power the input transducer. As mentioned in Chap. 1, it is highly desirable to use the same reference for the entire circuit. When a single reference is used, there will be automatic temperature/time tracking for critical portions of the circuit. Ratiometric measurements like this place little demand on the tolerances of the reference (with the exception of possibly noise). Absolute measurements, however, demand that the reference tolerances stay within the desired accuracy over the entire operating temperature range. To aid in selecting a reference for your design, some specifications typically stand out and are listed below.

Important reference specifications include:

- Initial tolerance, %
- Drift, ppm/°C, or %/°C
- Noise, nV/$\sqrt{\text{Hz}}$
- Reference voltage output
- Output impedance/drive

When a reference is considered for absolute measurement systems where it is undesirable to perform calibration, the reference must have an initial tolerance to match the desired accuracy. This can prove to be an expensive, or even impossible, approach for high-accuracy designs. The following list of acceptable tolerances of the reference for a given A/D resolution will help bring this point home.

A/D resolution, bits	Maximum tolerance, %
6	1.5
8	0.4
10	0.1
12	0.024
14	0.006
16	0.0015
18	0.00038
20	0.0001

Although some A/D converters come with an internal reference (i.e., bandgap), most precision A/D systems require an external reference. Absolute measurements greater than 8 bits generally fall into that category. This is often due to the incompatibility of the process used for the A/D converter to create an accurate and, most important, stable reference. For applications requiring only resolution, or ratiometric measurements (not absolute accuracy), the internal references can do the job. However, keep in mind that even when absolute measurements are not required, high-resolution measurements will very likely demand a separate low-noise circuit.

Some low-resolution absolute measurement applications even allow for the use of the A/D supply voltage for the reference as well. This is possible only when the tolerance of the supply over the full operating range (temperature and load) is within the required accuracy. When this low-cost approach is considered, great care must be taken to isolate the noise on the supply line from entering the A/D converter.

This will require careful layout (see Chap. 9) and filtering of the supply lines.

Bandgap references

References can be divided into two basic types: bandgap or zener-based. Bandgap references rely on the basic principle of the energy bandgap of a semiconductor material (1.205 V for silicon at 0 K) as well as other circuit components for load current and temperature compensation. References such as the National Semiconductor LM385 utilize this method. Monolithic CMOS A/D circuits commonly employ bandgap references as well. From a performance standpoint, bandgap references have higher noise than zener-based references (1000 versus 200 $nV/Hz^{1/2}$ and generally lower stability over temperature and time. Also consider that the bandgap reference produces a lower voltage (1.2 V) than the A/D converter typically calls for, so the bandgap reference must be amplified. What this means is that the actual signal-to-noise ratio will be even larger compared to the buried zener type that is typically higher in voltage. For these reasons, bandgap types are usually used only in low- to moderate-resolution CMOS integrated solutions.

Zener references

Zener-type references are available as discrete, as shown in Fig. 3.1*a*, or as part of a reference integrated circuit (that is, LM369 in Fig 3.1*b* from National Semiconductor). All diodes exhibit a point where a breakdown will occur due to reverse voltage bias. The buried-type zener (as opposed to the surface type) has very well-controlled characteristics. Discrete zeners offer a very low-cost reference compared to

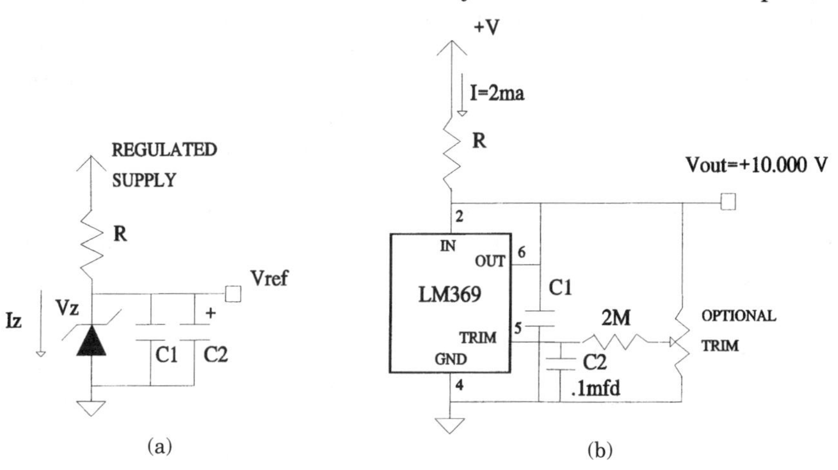

Figure 3.1 Common reference circuits. (*a*) Discrete zener diode; (*b*) integrated-circuit zener diode based.

an integrated solution; however, extra circuitry is required to achieve the same performance.

Ratiometric or low-resolution absolute measurements (<1 percent accuracy) can take advantage of the common discrete zener reference. Zeners are available in a variety of voltages, current levels, temperature drift, power ratings, and noise levels. Various operating voltages are possible by varying the amount of diffusion which correspondingly alters the voltage breakdown values. Zeners are available with voltages from as low as 3.3 V to very high if desired (>100 V).

General-purpose zener diodes below 5.6 V will have a negative temperature coefficient (TC), whereas zeners above 5.6 V will have a positive TC. Fortunately, one of the best (uncompensated) values is at 5.1 V (± 0.03 percent per degree Celsius) which is commonly used for most logic circuits. However, for wide temperature ranges even this can prove unacceptable. As an example, consider the following conditions:

Temperature drift example for 100°C range:

$$\text{TC of 1N5231B} = \pm\, 0.03\% \text{ at } I_z = 20 \text{ mA}$$

$$\text{Maximum error} = 100 \times 0.03\% = 3\%$$

Keep in mind that the tolerance and temperature drift of these discrete zeners are specified only for a given reverse current. Any deviation from this current will mean that the tolerance will widen.

For absolute measurements over a wide temperature range, the temperature-compensated zener can make a good choice. The available range of temperature-compensated zeners is from 6.2 to 11.7 V with a temperature drift from 0.01 to as low as 0.0005 percent. Below is a partial list of some temperature-compensated zeners from Motorola.

One of the potential problems with using temperature-compensated zeners is that they are available only in voltages above 6.2 V. This usually means that a resistor divider along with a buffer amplifier circuit must be used for typical A/D input ranges. Always take into account that temperature-compensated zeners are also characterized at a specific operating current. Depending on the delta-voltage ($V_{supply} - V_{ref}$) changes, a regulated zener supply may be required to maintain the zener current.

Temperature-Compensated Zeners

Type	V_z, V	I_z, mA	Drift, %/°C
1N825	6.2	7.5	0.002
1N4567	6.4	0.5	0.002
1N943	11.7	7.7	0.002

Another potential problem with discrete zeners in general is the amount of noise generated. All zeners will generate noise when biased in the forward zener direction. The largest part of the noise is from the zener breakdown phenomenon called *microplasma.* This is considered to be white noise with a constant amplitude versus frequency. To a lesser extent, the dynamic internal impedance of the zener will cause noise to be produced. For instance, if the supply line has frequency components, the zener current changes times the dynamic resistance will look like noise to the A/D reference input.

Noise from the internal zener impedance can be rather easily controlled by selecting the proper shunt capacitors across the zener. If there is concern about the zener reference noise, a "low-noise" zener such as the Motorola 1N4625 should be considered. This class of zener will typically generate a noise density of 250 $\mu V/Hz^{1/2}$. To keep the noise level to a minimum, be aware that as the zener current increases, the noise level decreases. Conversely, increased temperatures will increase the noise level. Attenuating the white noise will likely require a large resistor with a capacitor to form a low-pass filter. Therefore, a buffer amplifier is probably a good idea for minimizing the impedance seen by the A/D reference input.

Integrated references

The integrated solutions, such as the LM369 mentioned above, are much easier to design with, and they provide higher performance than discrete zeners. Integrated reference circuits are typically available with fixed outputs of 2.5, 5, and 10 V. The advantages include much greater accuracy (better than 0.1 percent) with a trim pin for additional accuracy as well as insensitivity to changes in the supply voltage. Another major advantage is their ability to sink and source current. This will prevent voltage spikes from input circuits to A/D converters. There are several points to consider, however, in order to maximize performance. The most important is trim pin 5. By placing a capacitor from pin 5 to ground, the noise level at the output can be reduced substantially. With $C_2 = 0.1\ \mu F$, the output noise is reduced from a flat response at about 20 μV rms to about 5 μV with a low-frequency roll-off. Here are some additional points to consider in designing with an integrated solution like the LM369:

1. Minimize leakage at pins 1, 3, 7, 8, and especially pin 5, by using guard rings connected to a low-impedance source. This will prevent the output from drifting.
2. Adjusting the trim pin can cause output temperature degradation (about 1 ppm for every 30-mV adjustment).

3. Internal power dissipation should be kept to a minimum (can add approximately 4 ppm/100 mW).
4. Preloading the output with a light load can prevent the class B type of amplifier from causing voltage spikes during load current polarity changes (i.e., sampling type A/D converters).
5. Check the data sheet for the recommended capacitor for output stability. The LM369 requires a tantalum capacitor or a 0.1-μF capacitor with a series 10-Ω resistor.

Integrated reference noise considerations

Regardless of the type of reference chosen, or whether the system is absolute or ratiometric, the noise level generated should be noted. There would be little value in using a high-resolution A/D converter along with a reference circuit that generates several least significant bits (LSBs) of noise. For high-resolution and precision needs, the circuit shown in Fig. 3.2 can be used to filter out much of the wideband reference-generated noise. The 10-kΩ resistor and 10-μF capacitor (tantalum) on the output of the reference perform a low-pass filter with 3-dB roll-off at 1.6 Hz:

$$f(3\text{ dB}) = \frac{1}{RC(6.28)} = \frac{1}{(10\text{ k}\Omega)(10\text{ }\mu\text{F})(6.28)} = 1.6\text{ Hz}$$

Since the reference inputs to A/D converters typically require low impedance, an amplifier buffer should be used. This amplifier needs to be a low-noise type such as the OP-27, which produces only about 3 nV/Hz$^{1/2}$ of noise. If further noise reduction is required, the output of the amplifier can be filtered with the 100-Ω resistor and 10-μF capacitor. This forms a second low-pass filter with a 3-dB roll-off at 160 Hz. To avoid oscillation problems with the heavy capacitive load (10 μF), the standard isolation technique can be employed (R_0, R_f, C_f). Again,

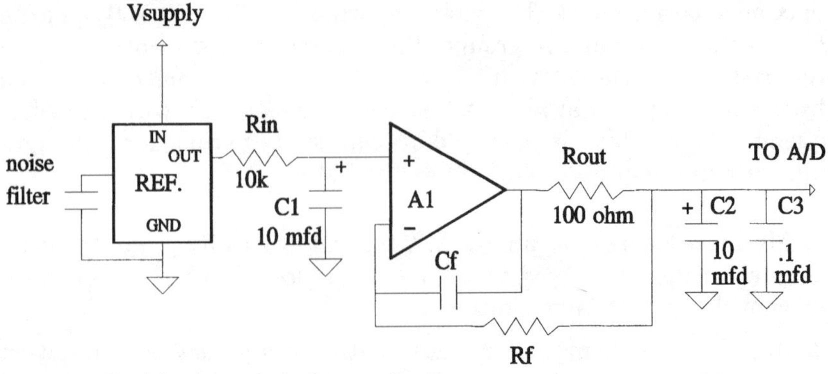

Figure 3.2 Precision low-noise reference circuit.

as standard procedure, the 0.1-µF ceramic capacitor (C_3) should be used in parallel with the 10-µF capacitor (C_2) to reduce the high-frequency series impedance of an electrolytic alone.

Integrated reference temperature drift

Very high-accuracy A/D converter designs that must operate over even a limited temperature range must have some type of compensation. As an example, consider a 16-bit system with a 5-V reference. This will require the reference temperature drift over the operating range to stay less than 5/(65, 536) V, or 76 µV. Assume that we are dealing with a 70°C temperature swing. Then the required reference ppm/°C drift to maintain 16-bit accuracy is

$$\Delta V_{ref} = 76\ \mu\text{V} = \frac{5\ \text{V}\ (\text{ppm}\ /^\circ\text{C})(70^\circ\text{C})}{10^6}$$

Solving for ppm/°C: $\text{ppm}/^\circ\text{C} = 0.2\ !$

High-quality references provide only about 3 ppm/°C. This means that some form of temperature compensation will be required. There are numerous techniques for accomplishing this, but all share some similarity. The concept is based on measuring the reference ambient temperature and then generating a correction factor from known reference characteristics. One option can be to digitally correct each measurement with the system processor, which can measure the temperature with the A/D converter, and perform a mathematical correction on every conversion. This will, of course, take instruction time and result in longer system throughput.

Another option, for faster conversion cycle time, is to utilize an automatic approach in hardware, as shown in Fig. 3.3. The circuit will

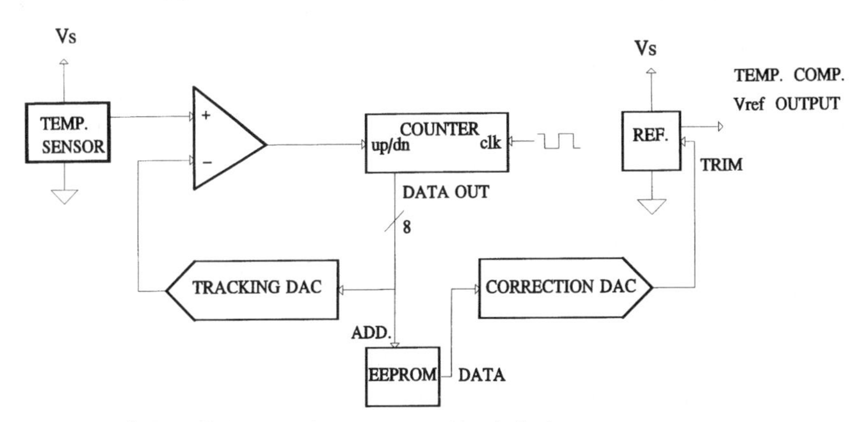

Figure 3.3 Automatic temperature compensation technique.

automatically monitor the reference temperature by using a tracking A/D converter technique. This technique uses a free-running counter that continuously counts up or down, depending on the comparator output. Counter outputs then drive the measurement digital-to-analog converter (DAC) for producing an analog signal to match the temperature sensor. In addition, the counter output also addresses an EEPROM where the corresponding correction coefficients are stored. Finally, the EEPROM drives a correction DAC for generating the final analog output correction for the adjustable reference. Typically, the correction coefficients will be determined during an autocalibration cycle. This can be done in the circuit with the EEPROM.

Operational Amplifiers

An operational amplifier, or op amp, is generally required to scale and filter the raw analog signal for the A/D converter. Before an op amp is chosen, however, it is necessary to determine the important specifications. Basically we need to decide if the largest concern is for good dc or ac specifications. For instance, systems such as Digital Audio require very high ac performance. This includes low total harmonic distortion (THD) and high gain bandwidth. THD is largely caused by loss of gain at higher frequencies, which causes an increase in V_{os} for the op amp. This results in an increase in nonlinearity, which is the real cause of distortion. Important dc and ac specifications are listed here:

Key dc specifications:

1. Input offset voltage V_{os}
2. V_{os} temperature drift
3. Input bias offset current I_{bos}
4. I_{bos} temperature drift
5. Open-loop gain A_v
6. Open-loop gain temperature drift
7. Noise
8. Common-mode rejection ratio (CMRR)

Key ac specifications:

1. Bandwidth (gain bandwidth)
2. Total harmonic distortion (THD)

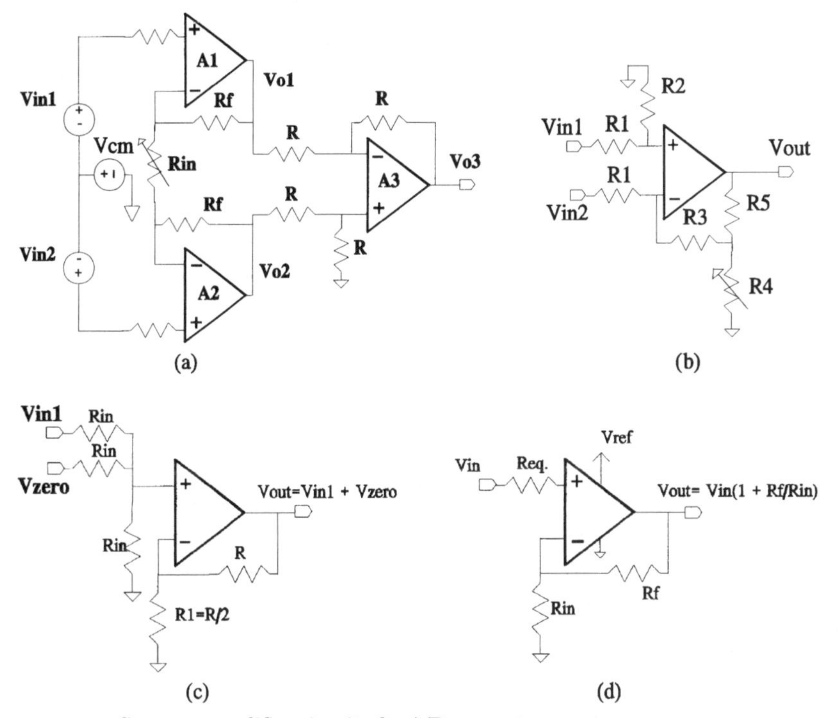

Figure 3.4 Common amplifier circuits for A/D converters.

3. Slew rate (V/µs)
4. Settling time
5. Output current
6. Noise

In the successive illustrations of Fig. 3.4, we see some of the useful amplifier circuits that can be used in an A/D system. The design engineer has many choices of the type of circuit and type of op amp. This includes CMOS, bipolar, BiFET, chopper-stabilized, voltage/current feedback, and BiCMOS amplifiers. For obvious reasons, not all of the options can be discussed in this book, and some are left to dedicated books on the subject.

Classical instrumentation amplifier

Figure 3.4*a* is an example of the classical instrumentation amplifier. Three amplifiers are used to create a "near ideal" gain stage with very high input impedance (amplifiers A_1 and A_2). Amplifiers A_1 and A_2 also provide the differential gain:

$$V_{o1} = V_{\text{in1}}\left(1 + \frac{R_f}{R_{\text{in}}}\right) - \frac{V_{\text{in 2}}R_f}{R_{\text{in}}} + V_{\text{cm}}$$

$$V_{o2} = V_{\text{in2}}\left(\frac{1 + R_f}{R_{\text{in}}}\right) - \frac{V_{\text{in1}}R_f}{R_{\text{in}}} + V_{\text{cm}}$$

Assuming that output amplifier A_3 has matched resistance values ($R_1 = R_2 = R_3 = R_4$), the output V_{o3} will equal $V_{o2} - V_{o1}$ and can be expressed as

$$V_{o3} = (V_{\text{in2}} - V_{\text{in1}})\left(1 + \frac{R_f}{R_{\text{in}}}\right)$$

Usually the circuit gain is provided in the front end by varying R_{in}. By sharing R_{in} between A_1 and A_2, it becomes much easier to change gains and still match the ratio of R_f/R_{in} for both A_1 and A_2. When you are designing any differential amplifier, it is important to realize the effects of any mismatch in the circuit resistance ratios. Mismatch in resistance will show up in the output as common-mode rejection ratio (CMRR) error. For example, if $V_{\text{cm}} = 10$ V and $V_{\text{in}} = 1$ V span, a 0.1 percent error in matching the ratios R_2/R_1 will result in an error of

$$\text{Output error} = \frac{(0.1\%)(10\text{ V})}{1\text{ V}} = 1\%$$

Single-point gain differential amplifier

When a single-stage differential amplifier is used and it is desirable to provide high gain, a problem can quickly arise. For high gain, the feedback resistance will have to be extremely large. The reason is that the input resistance will have to be lower than the feedback value, yet large enough not to load down the input signal. The input resistance will also have to be large enough to prevent the gain from being affected appreciably by changes in the input impedances. Figure 3.4*b* is an example of a single-point gain adjustment. One value of this idea is that the gain can be adjusted with just one resistor (R_4). Additionally, the circuit resistance can be reasonably low, with the exception of R_{in}. The desirable high input impedance from R_{in} is possible because of the multiplying effect of the feedback divider resistors (R_4 and R_5). The individual resistance values can be calculated as shown:

- Set $R_4 << R_5 << R_3$.
- Let R_{eq} = parallel combination of R_4 and R_5 (use R_4 nominal value).

$$R_f = R_3 + R_{\text{eq}}$$

- Solve for $R_f = R_2$ (for common-mode rejection).

$$V_{gain} = \frac{R_3(R_4+R_5)}{R_1R_4}$$

Noninverting summing amplifier

Figure 3.4*c* shows a noninverting summing amplifier that can work off a single supply and is handy for providing an offset for an A/D measurement. The trick is to make the impedance seen by the input signals the same and to set the gain to compensate for the input resistor divider. For best performance, a resistor network of equal values works very well for matching and tracking purposes. If more inputs need to be added, simply change the value for R_1 to equal the lower portion of the input resistor divider, as shown:

- Let R_{eq} = parallel combination of $n - 1$ input resistors.
- Set $R_1 = R_{eq}$.

Rail-to-rail amplifiers

For simple input signals that require no offsets and only a gain for the A/D converter, the circuit in Fig. 3.4*d* will suffice. With CMOS amplifiers (i.e., National Semiconductor LMC6xxx series) it is possible to have both rail-to-rail input range and, more importantly, output swing. This means that a low-level input signal can be scaled to equal the A/D input range by an amplifier powered by the same supply level as the A/D converter.

Adjustable-gain instrumentation (two op amps)

Figure 3.5 shows an adjustable-gain version of the instrumentation amplifier with two op amps. This circuit works well for providing high input impedance to the signal and saves on one amplifier over the classical circuit. The circuit basically works by double-amplifying the negative

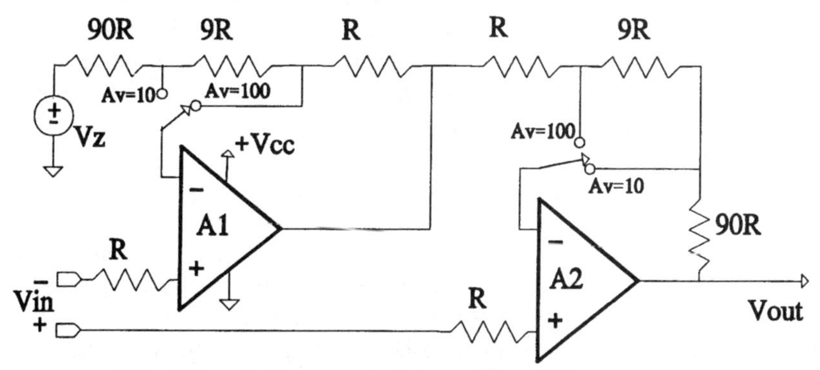

Figure 3.5 Adjustable-gain instrumentation amplifier with two op amps.

input so that its total gain is unity at the output. Amplifier A_2 then provides the differential gain of V_{in}. Voltage V_z (if not ground) will cause the output to be biased by the same voltage. If the amplifier output swing is a problem, V_z can be modified to ensure operation within the amplifier's input/output range for maintaining circuit linearity. This allows for a wider selection of op amps to be used with a single supply. The neat thing about this circuit is that the gains can be altered without introducing errors from the analog switches. This is because the switches are in the high-impedance portion of the amplifier feedback loop.

Discrete instrumentation amplifier

It is important for the ratios of the resistances to match, but often it is more important that the ratios are stable over temperature and time. This is where a discrete instrumentation amplifier has an advantage over a fully integrated version. Processes used for integrated circuits do not allow for the resistances to be easily matched for temperature tracking. The initial tolerances can be easily laser-trimmed, but the temperature tracking can be a problem for high-accuracy designs. Another little-known problem with integrated-circuit resistors is that the voltage coefficient affects the resistor value. This is not nearly as prevalent with discrete resistors. Discrete designs utilizing custom or standard thin-film resistor networks, as discussed in Chap. 2, can offer many advantages, the most notable of which is the temperature tracking down to under 1 ppm/°C if desired.

Single-supply amplifiers

Single-supply amplifier operation can offer some cost savings, but it requires some special attention. There are many amplifiers available that operate with a single supply, and most include ground within the linear input range. Depending on how the amplifier is configured, you may have problems, however. The most likely problem area is in the output swing. If the output does not swing very near ground, the net A/D input range will be limited. A/D converters with both a minus and a plus reference input can eliminate the problem with limited amplifier output swing. For example, the two op amp instrumentation amplifier or the noninverting summing circuit can use V_z for the negative A/D reference. Thus, a 0-V input will still result in zero A/D counts.

Regardless of the chosen amplifier, the actual voltage output drops will depend on the amount of load current. To a lesser extent, the limited input ranges of most amplifiers can also be troublesome. Most amplifier inputs only go to within about 1.5 V of the positive supply rail. This may not be a problem in most applications since the input can be divided with resistors if it is too large. These problems are largely avoided when the amplifier has both rail-to-rail input and output capability, as with the National Semiconductor LMC6484 type (providing speed is not an issue).

High-Speed Amplifiers and Buffers

High-speed A/D converters measuring high-frequency input signals require an amplifier with special characteristics. This applies to flash, subranging, and successive-approximation types of converters. Although each will have somewhat different needs, these applications generally require an amplifier with wide bandwidth and low distortion. To avoid distortion, it is recommended that the amplifier full-power bandwidth (FPBW) be approximately four times the maximum frequency, calculated by

$$\text{Maximum frequency} = \frac{\text{slew rate}}{6.28V_{\text{p-p}}}$$

where slew rate = output capability, V/μs
$V_{\text{p-p}}$ = output peak-to-peak swing, V

For conversions not to be limited by speed, the A/D input capacitance must be charged or discharged very quickly. This will require an amplifier or buffer with high output current that can charge the internal capacitance and meet the settling time requirement.

One potential problem with wide-bandwidth amplifiers is their susceptibility to oscillation due to capacitive loading. Typically this capacitance only needs to be in the range of 10 to 20 pF to be a problem. To compensate for this, a small series output resistor of 5 to 50 Ω may be all that is necessary to isolate the capacitive load. Other applications may need to have the standard capacitance compensation circuit and power supply filters as shown in Fig. 3.6. The manufacturer's data sheet and application notes should be consulted to ensure proper compensation without performance degradation.

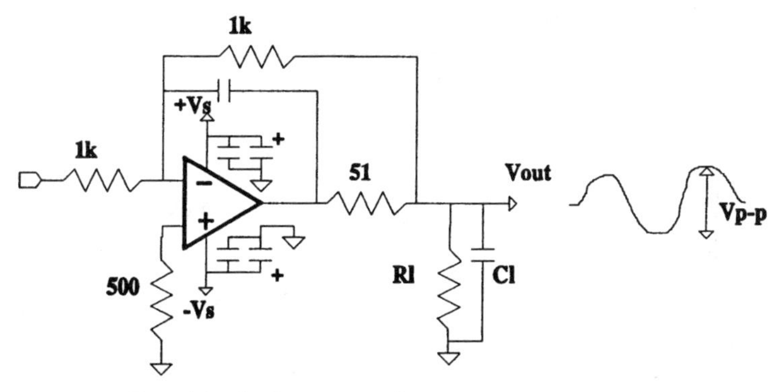

Figure 3.6 Capacitive load compensation.

Current feedback amplifiers

Current feedback amplifiers have an advantage over voltage feedback amplifiers when the application calls for high gain of a high-frequency input signal. The negative input of current feedback amplifiers consists of a low-impedance coupled emitter stage. This differs from the conventional voltage amplifiers that have both negative and positive high input impedance. For this reason, current feedback amplifiers have the characteristic of relatively constant gain versus bandwidth. Conversely, voltage feedback amplifiers have a gain-bandwidth limitation where the bandwidth will very quickly start rolling off with increasing gain. For practical purposes, gain bandwidth does not apply to current feedback amplifiers. Of course, there is a limit to the available bandwidth. This is determined by the value of the feedback resistor which sets the closed-loop dominant pole location. Increasing the feedback resistor will increase the gain; however, it will also decrease the bandwidth. Another tradeoff to consider is that decreasing the feedback resistor will cause less phase margin.

Buffer amplifiers

Buffer amplifiers are attractive for their high speed and high-output-current characteristics. These devices are often configured as voltage followers with just one input, and typically they have a voltage gain of about 0.97 that will vary with the load. The poor direct-current (dc) specifications may seem to be a real problem, but usually there is a dc correction built into the system to compensate for the total dc errors. Buffer amplifiers typically find their way into flash converter applications. Buffer amplifiers can also be placed in the feedback loop of slower amplifiers with less drive to boost the output current and slew rate (Fig. 3.7). In this application, the dc gain and offset error from A_2 would be automatically compensated for by A_1 since A_2 is in the feedback loop. The total open-loop gain of the circuit A_{vol} is simply

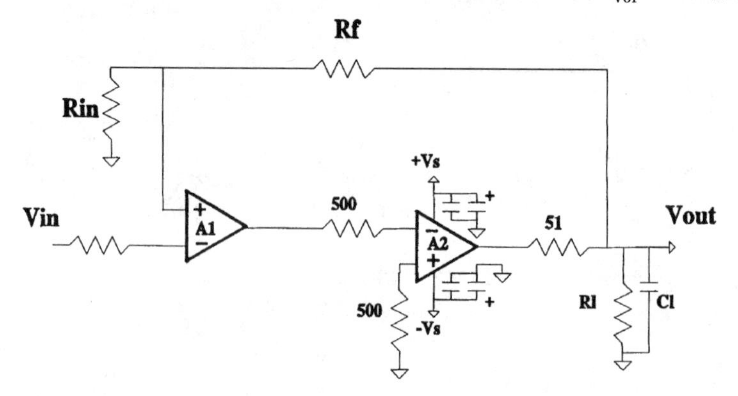

Figure 3.7 Bandwidth improvement with a buffer amplifier.

$$A_{vol} = |A_{vol1}|\,|A_{vol2}|$$

With A_2 having the larger slew rate, the A_1 output signal will be quite small in comparison. This relaxes the bandwidth requirements of A_1 significantly. However, the composite arrangement of the two amplifiers will cause additional phase shift. For instance, if each amplifier has a single (20-dB) pole at frequency f_c, then the composite response will be the product of each. In other words, there will be a − 40-dB roll-off starting at f_c.

Comparators

Although not a glamorous device, the comparator performs a vital function in all A/D converters. For example, flash converters utilize several comparators (that is, 256 for an 8-bit conversion), and sampling converters require a comparator for making a decision with the successive-approximation register. The basic difference between a comparator and an amplifier is that the comparator does not have to be compensated for closed-loop feedback. This means that the internal compensation capacitor is not needed, and this results in an increased output slew rate.

Comparators are characterized by their voltage gain in V_{out}/V_{in} (units of volts per millivolt) and slew rate (units of volts per microsecond) for a given "overdrive." *Overdrive* refers to the amount of differential voltage at the input pins, and generally it has a significant effect on the slew rate. In typical A/D converter applications, the comparator must slew its output quickly, and without oscillation once the input thresholds are crossed. Since even the slightest amount of noise present in the system could cause output oscillations during a threshold crossing, some precautions should be taken. To optimize performance, the following options ought to be considered.

Optimizing comparator design:

1. Add hysteresis (positive feedback typically 10 to 20mV)
2. Separate the high-speed output traces from the input signals.
3. Use bypass capacitors very close to supply pins.
4. Keep input resistance low and place close to input pins.
5. Use guard ring around input pins and guard plane under input and output pins.
6. Place small-value capacitor (approximately 100 to 1000 pF) between the input pins (shown as option in Fig. 3.8) to assist in making a smooth output transition during threshold crossing.

The circuit in Fig. 3.8 shows hysteresis (positive feedback), which helps in two ways. First, by changing the input threshold by some finite

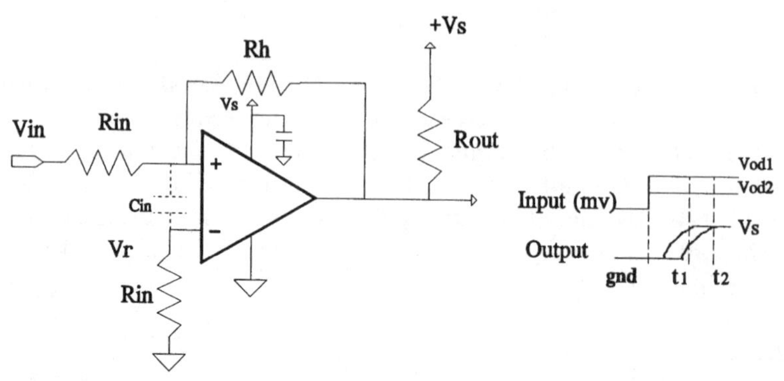

Figure 3.8 Comparator employing hysteresis.

amount during the output swing, the noise immunity is improved. Second, with more overdrive at the moment of threshold crossing, the slew rate will increase. This provides the benefit of getting the input through the threshold region much faster (thus helping to avoid system noise at the input). The amount of hysteresis V_h in Fig. 3.8 can be calculated by

$$V_h = (V_s - V_r)\left(\frac{R_{\text{in}}}{R_{\text{in}} + R_h}\right)$$

where V_r = comparator trip point; zero volts in this equation (assuming amplifier $R_{\text{out}} << R_h$).

Sample-and-Hold

The purpose of the sample-and-hold (S/H) is to track and hold the input signal long enough for the A/D converter to complete a conversion without appreciable error. When a S/H is not included internal to the A/D converter, an external sample-and-hold is required if the input signal can change by more than ½ LSB during a conversion cycle. This can be due to either the actual input signal frequency or system-induced noise causing the input to change rapidly. Subranging and successive-approximation A/D converters are the largest applications where sample-and-holds are found. Typical A/D converters that do not require a S/H include integrating, PWM, and delta-sigma converters. These converters are not susceptible to high-frequency signals (or noise) due to their averaging mode of operation.

Besides high signal frequencies, there are other possible reasons for choosing a S/H. For example, peak detectors operate by capturing fast input signal peaks on a holding capacitor so the A/D converter can measure the signal over a longer period. Some applications benefit by using a S/H to synchronize several channels. For instance, eight channels can be simultaneously held and then sequentially measured by the A/D converter. Other applications take advantage of sample-and-holds for reducing crosstalk in the multiplexer. Some of the important characteristics for selecting a S/H are listed here.

S/H important characteristics:

1. Input/output offset voltages
2. Charge injection (pedestal error)
3. Drift (capacitor droop)
4. Gain error
5. Bandwidth limitation (slew rate, acquisition time T_a)
6. Distortion

To ensure maximum system performance, it is very important to understand the operation of various S/H architectures and the tradeoffs associated with each. As the operating bandwidth and/or the dc accuracy requirements increase, so does the importance of the S/H. In these high-performance applications, it is likely the S/H will be the limiting factor in overall performance (i.e., bandwidth, distortion, dc accuracy).

Open- and closed-loop S/H architectures

There are two basic types of S/H architectures: open- and closed-loop (shown in Fig. 3.9). The basic ideas are the same for both in that a switch is required to charge a capacitor for holding the input voltage long enough for a complete conversion. To avoid long *RC* delays from the total input resistance (R_{in}+R_{switch}) with the holding capacitor, and loading (droop) on the hold capacitor, an input/output buffer is also required. Open-loop-type sample-and-holds are faster than closed-loop types which have the delayed output fed back to the input buffer. However, closed-loop architectures provide higher dc accuracy because of this feedback (output amplifier offset errors cancel).

Charge injection

Charge injection can occur during the transition between the sample-and-hold modes. When the S/H is configured as a voltage follower, as in Fig. 3.9*a* and *b,* the amount of charge injection, or pedestal voltage, will be proportional to the input voltage. The integrator-type configuration (Fig. 3.9*c*) does not have the input voltage dependency since the input amplifier does not drive the holding capacitor directly. When the open-loop S/H switch is opened, a charge is transferred to the holding capacitor, which develops a pedestal voltage V_p equal to

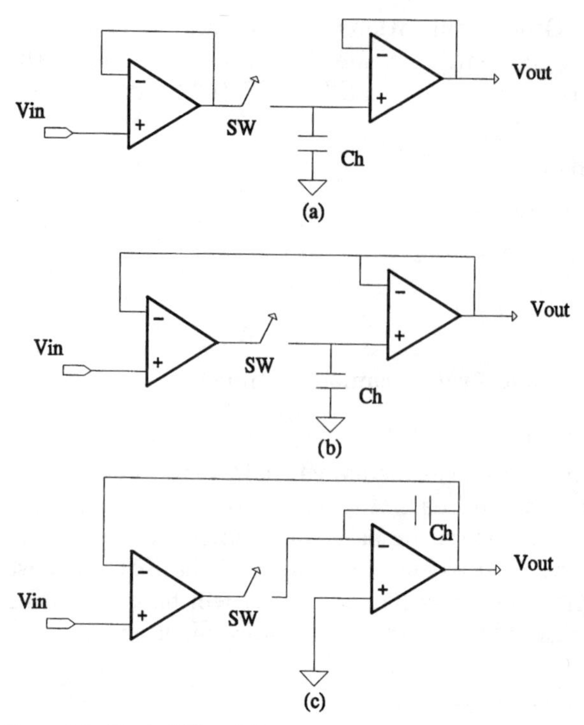

Figure 3.9 Basic S/H architectures.

$$V_p = \frac{(V_c - V_{in})C_s(V_{in})}{C_h}$$

where V_c = total clock voltage swing
$C_s(V_{in})$ = switch capacitance as function of input voltage
C_h = holding capacitance

Closed-loop sample-and-holds like that shown in Fig. 3.7*c* charge the holding capacitor within the relatively high-impedance feedback loop of the output amplifier. Because the summing input of A_2 remains at virtual ground, the charge injection voltage dependency will be significantly reduced. The amount of pedestal voltage with this architecture is

$$V_p = \frac{V_c C_s}{C_h}$$

It is common to specify the magnitude of charge injection in units of picocoulombs (pC) with the pedestal voltage computed from the basic equation of V = charge/capacitance:

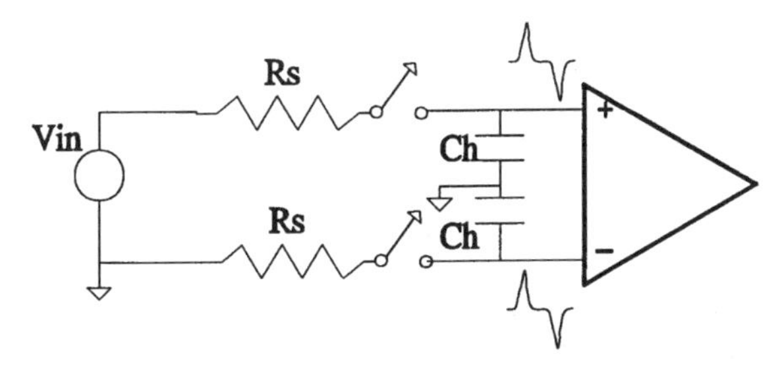

Figure 3.10 Dual-switch charge injection cancellation.

$$V_p = \frac{\text{charge, (pC)}}{\text{holding capacitance, pF}}$$

In applications where the pedestal error is unacceptable, there are techniques that can be applied to greatly reduce the effects. One such technique is to use a dual-switch action for a cancellation effect, as shown in Fig. 3.10. By providing equal amounts of charge transfer in both inputs of the differential amplifier, the net error will be drastically reduced. This same technique is commonly applied in the delta-sigma A/D converters or any highly sensitive switching application.

Pedestal error can be compensated for in another way. Sample-and-holds that utilize the closed-loop architecture and the holding capacitor within the output stage allow for hardware trimming. In other words, if the input signal magnitude does not affect the amount of charge injection, the error can be considered a constant offset and can be adjusted out.

Current multiplexer S/H

By using the same dual-switch technique described above, the current multiplexed architecture in the LF6397 from National Semiconductor adds some performance. It accomplishes this by combining the speed associated with the open-loop approach with the accuracy of the closed-loop method. Operation of the circuit in Fig. 3.11 is as follows. In the track mode, S_1 connects the transconductance amplifier A_1 to the buffer A_3 in a closed-loop mode. During this time, hold capacitor (C_h) is charged to V_{in}, and the dummy capacitor C_d is discharged with S_2 and S_3 closed. With the application of the hold command, all the switches will change state. This now connects amplifier A_2 to the output amplifier A_3 in an open-loop manner. At the moment of the switch transition, an equal amount of charge is added to the dummy capacitor and to the holding capacitor. Thus the charge injection is canceled with differential amplifier A_2.

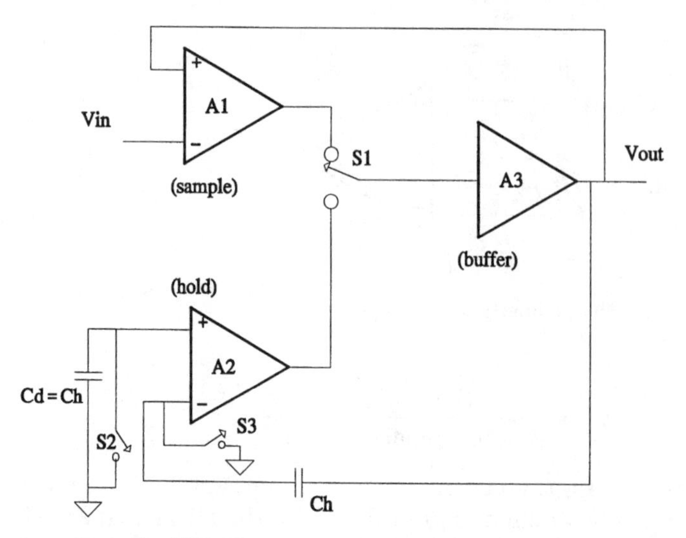

Figure 3.11 The LF6397 current multiplex S/H architecture.

S/H capacitor selection

Hold capacitors (C_h) are selected based on the need to minimize the amount of droop V_d within the A/D conversion time. If the internal capacitance of the S/H is inadequate, the value may have to be increased. We say *may have to be increased* because this will increase the S/H acquisition time and may become a problem if it is too large. This is due to the capacitor forming a low-pass filter in combination with the input buffer output resistance and switch resistor. In addition, the input buffer amplifier must have sufficient output current to drive the capacitance within the required time. An acceptable holding capacitor should provide minimal leakage current and have low dielectric absorption. With all this in mind, the required capacitance value can be easily calculated from the following:

$$C_h = \frac{I_L\,dt}{V_d}$$

where I_L = total leakage current of switch and output amplifier over temperature

dt = A/D conversion time

V_d = acceptable capacitor droop (that is, ½ LSB)

As mentioned above, the hold capacitor, if it is too large, can affect the S/H acquisition time. If this is the case, then another S/H design should be considered. Either a S/H with reduced leakage current (output buffer amplifier input leakage and switch leakage) for the holding capacitor, or a buffer with more drive current should be considered. To determine if the S/H acquisition time is adequate, the following calculation can be made:

$$T_a = \frac{C_h V_s}{I_{dr}} + t_s$$

where V_s = full-scale step voltage
T_a = total S/H acquisition time
I_{dr} = drive current for holding capacitor C_h
t_s = settling time of output amplifier for required full-scale step voltage

Building a discrete S/H

In some applications it may pay to build a lower-cost discrete S/H compared to a monolithic or high-performance hybrid. When you undertake such a task, several requirements must be met. First, the A/D conversion time will largely determine the required acquisition time of the S/H. This will set the switch-on resistance and hold capacitor limit values to allow for the input signal to be acquired within the desired time frame. Switch turn-on and turn-off times are also factors that can affect the throughput rate of the A/D converter. All sample-and-holds require an input buffer that can drive the hold capacitor quickly and an output amplifier with high input impedance. To keep the total droop on the holding capacitor less than ½ LSB during a conversion, the switch and output amplifier leakage must be kept very low.

As an example, Fig. 3.12 shows a discrete S/H that is capable of supporting a 12-bit A/D converter with approximately 1-µs acquisition time of a ±5-V step. Note that this is a closed-loop architecture with an integrating output stage. This means that the dc accuracy will be favored at the expense of some speed. The circuit operates by first closing S_1-*A*, S_2, and S_3, with S_1-*B* open. This provides a charge

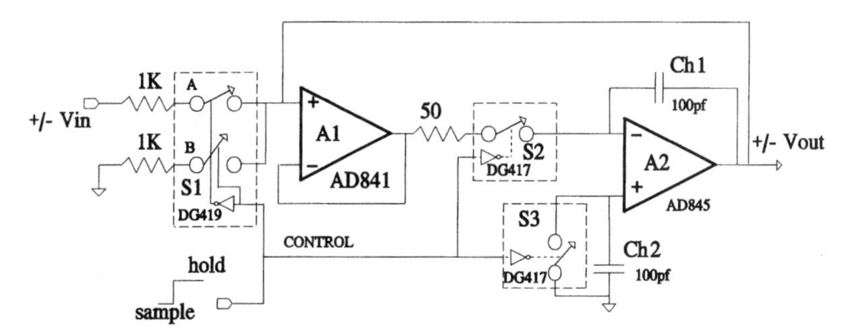

Figure 3.12 Low-cost discrete S/H.

path for the holding capacitor (C_{h1}) in the feedback of A_2 while in the sample mode (logic low). After the signal is acquired, all the switches change state, with the control signal going logic high, thus holding the input signal on C_{h1}.

Amplifier A_1 (AD841) provides a buffer for the input signal to drive the holding capacitor with 50 mA. Other important characteristics of this amplifier are unity gain stable, high bandwidth (5 MHz), and capability of driving a highly capacitive load while maintaining stability. Amplifier A_2 (AD845) holds the sampled input signal and drives the A/D input stage. This amplifier provides low input leakage current (500 pA) with a FET input stage and can settle to within 0.01 percent of the final output value in less than 250 ns.

Key requirements of the CMOS switches used in this example (DG417/419) are their high switching speed of 100 ns and low on resistance of 20 Ω. Other key requirements include the switch operating voltage range of ±15 V, which allows for wide input range and easy logic level control. Among the disadvantages of the integrated switches compared to discrete FET switches (i.e., Philips SD5001) is their slower speed and higher charge injection. To meet the 12-bit accuracy requirement, a charge injection cancellation technique is used. The DG417 CMOS analog switches from Siliconix (S_2 and S_3) provide equal charge injection onto holding capacitors C_{h1} and C_{h2} for differential amplifier A_2. Another potential problem—feedthrough error—is essentially reduced to zero since switch S_1 disconnects the input during the hold mode and connects A_1 to ground.

Multiplexers and Analog Switches

Systems that have several signals to convert can often take advantage of the use of a multiplexer and/or an analog switch. Generally speaking, a multiplexer is just a version of an analog switch for sharing hardware with multiple signals. We discuss first the multiplexer and then applications for analog switches. When either is needed, there are common concerns:

Important characteristics:

1. Crosstalk/off isolation
2. On resistance
3. Input/output capacitance
4. Input/output leakage current
5. Switching speed (on/off, settling time)

6. Supply range
7. Break-before-make time

An example of a typical multiplexer application is shown in Fig. 1.1, where a common S/H, buffer, and A/D converter are utilized by all channels. As usual, nothing comes for free, and there are some tradeoffs in using a multiplexer, with the most severe disadvantage being crosstalk.

Crosstalk

Crosstalk is the amount of feedthrough from the multiplexer unselected or off channel(s) to the on channel. This parameter is frequency-dependent due to the capacitive coupling of the internal switches to the output. For this reason, it is commonly specified in decibels as a function of input frequency. Using a model of the multiplexer switches shown in Fig. 3.13 will help you understand how crosstalk is generated.

Crosstalk can be defined as consisting of three parts:

1. Static crosstalk
2. Dynamic crosstalk
3. Adjacent crosstalk

Static crosstalk is the ratio of input to output impedance while switches are in a static condition. For example, this can occur when an analog switch or multiplexer is used for setting the gain of a programmable amplifier. In this mode, it is the ratio of the C_{eq} impedance of the off channel(s) to the on resistance R_{on} of the selected channel that determines the level of crosstalk.

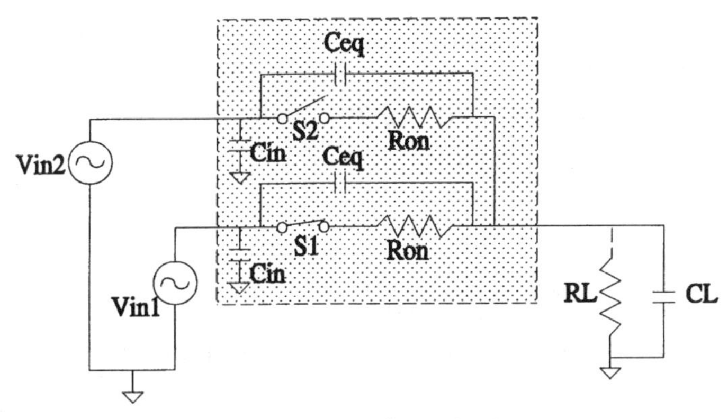

Figure 3.13 Equivalent circuit of analog multiplexer.

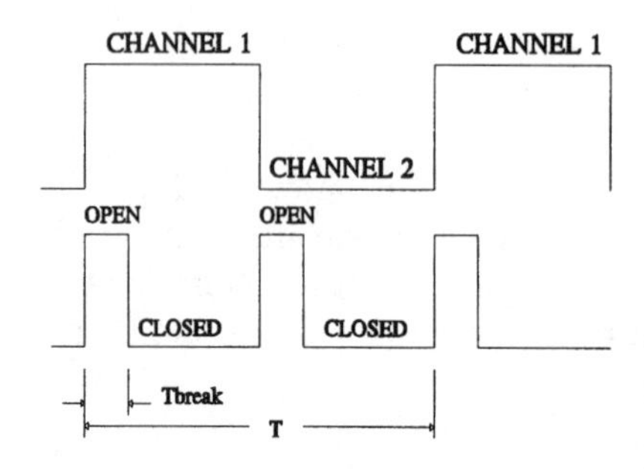

Figure 3.14 Approximation of multiplexer resistance waveforms.

Dynamic crosstalk occurs when analog switches are cycled through channels. This will be the case with a multiplexer used for selecting various input channels for the A/D converter. To determine the effective dynamic impedances, it is necessary to consider the frequency of operation, input signal frequencies, and break-before-make time. Referring to Fig. 3.14, we can compute the effective R_{eq} of the multiplexer by approximating the resistance as a function of the duty cycle. This will give an accurate result providing the load resistance R_L is much larger than R_{on}.

$$R_{eq} = \frac{R_{on}(\text{duty cycle 1}) + R_L(\text{duty cycle 2})}{T}$$

where duty cycle 1 = $T - 2T_{break}$
duty cycle 2 = $2T_{break}$
T = address clock period

Crosstalk calculation example. As an example, let's compute the amount of crosstalk for the DG406 16-channel multiplexer under the following conditions:

1. Unselected input frequency = 1 kHz
2. Channel select frequency = 20 kHz (or $T = 50\ \mu s$)
3. $T_{break} = T_{on} - T_{off} = 600\ \text{ns} - 300\ \text{ns} = 300\ \text{ns}$
4. $R_{on} = 120\ \Omega$
5. $R_L = 100\ \text{k}\Omega$
6. $C_{eq} = 8\ \text{pF}$ (source off capacitance $C_{s\text{-off}}$)

First, it is necessary to compute the equivalent resistance R_{eq} of the analog switches.

$$R_{\text{eq}} = \frac{120[50\ \mu\text{s} - 2(300\ \text{ns})] + 2(100\ \text{k}\Omega)(300\ \text{ns})}{50\ \mu\text{s}}$$

$$= 119 + 1200 = 1319\ \Omega$$

Next, the effective impedance of the off channel is computed by using the signal frequency and $C_{s\text{-off.}}$

$$X_c = \frac{1}{6.28(1\ \text{kHz})(8\ \text{pF})} = 20\ \text{M}\Omega$$

Last, compute $20 \log(X_c / R_{\text{eq}})$.

$$\text{Crosstalk} = 20 \log \frac{20\ \text{M}\Omega}{1319} = 84\ \text{dB}$$

Adjacent crosstalk is the third part of crosstalk to be concerned about. This parameter is not frequency-dependent; it is affected only by the dc levels and the order of switching on the multiplexer. The *adjacent channel* is the channel that follows the present on channel. Depending on the various adjacent channel signal levels and the charge/discharge time constants of C_L and R_x ($R_x = R_{\text{on}} + R_L$), this may not be a problem. However, a problem can arise when the load capacitance C_L does not have enough time to charge to its final value before the A/D converter samples it. By computing the time constants with R_x and considering the previous charge level, the time before a valid sample can be determined. This will be a problem only if the load capacitor is very large, the total charge resistance $R_{\text{on}} + R_{\text{in}}$ is large, or the sampling frequency is very high. In most situations, the multiplexer transition time and the settling time will provide an accurate estimate of the delay necessary between measuring multiple channels.

Crosstalk Elimination

Multiplexers can eliminate or significantly reduce dynamic crosstalk by taking advantage of the same technique used in sample-and-holds to eliminate feedthrough. By short-circuiting (or shunting) the input signal to ground during the hold mode, feedthrough crosstalk from unselected channels can be virtually eliminated. One example of this type of multiplexer is the Harris HI524, shown in Fig. 3.15.

Adjacent-channel crosstalk can be handled by alternately grounding input channels of a common multiplexer. This eliminates the effects of previous channels and makes the operation more predictable. Another method is to minimize the capacitive loading on the multiplexer out-

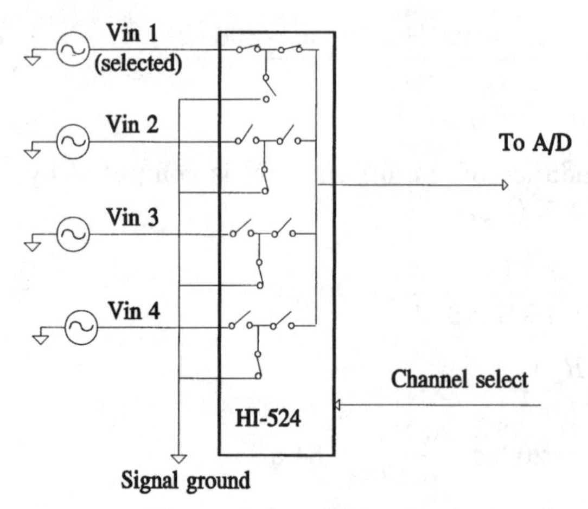

Figure 3.15 Shunt switch multiplexer.

put. To accomplish this, it is necessary to use careful layout and multiple tiering of multiplexers if necessary (as discussed in Chap. 1).

Determination of maximum system sampling rate

The on resistance of the switches can come into play under certain conditions that will tend to limit the sampling rate. Since there is always going to be some finite amount of output capacitance C_d, the on resistance R_{on} will form an *RC* low-pass filter. This will affect the settling time T_s of the switch and ultimately the maximum sampling rate of two consecutive channels. The maximum sampling frequency F_s for a multiplexer with n channels and transition time T_t can be determined from

$$F_s = \frac{1}{n(\mathrm{T_s} + T_t)}$$

where T_s = settling time based on desired bits of accuracy
$T_s = m(R_{on} \times C_{d\,on})$

For example:

m = 6 *RC* time constants (8 bits)
9 *RC* time constants (12 bits)

An example will help clarify the above. Using the DG406 16-channel multiplexer for a 12-bit A/D system, we determine the maximum sampling frequency.

$$F_s = \frac{1}{m(n \text{ channels})(R_{\text{on}})(C_{d\,\text{on}}) + \text{transition time}}$$

$$= \frac{1}{9(16)(100\ \Omega)(100\ \text{pF}) + 300\ \text{ps}}$$

$$= 694\ \text{kHz}$$

Since it is likely that the input filter will not have a perfect roll-off, and since the RC delays will be a factor, we need to add some margin for the maximum frequency. It is generally good practice to use a factor of about 4, so the maximum usable sampling frequency is

$$F_s = \frac{694\ \text{kHz}}{4} = 173\ \text{kHz}$$

Dealing with low-level signals

Differential multiplexers should be employed to deal with low-level signals. This will help reduce the effects of system noise by having balanced lines. For optimum performance, a separate S/H should be used for both input channels to a differential amplifier. It may also be advisable to use a buffer amplifier prior to the multiplexer to reduce the input leakage problems from either a low-level signal or high-impedance input. The most severe leakage will occur at higher temperatures and will appear as an offset voltage.

System Timing

The above analysis for determining the maximum allowable sampling frequency needs to be expanded if a multiplexer and S/H are used to feed the A/D converter. This will require that the total time for each stage be added. Figure 3.16 illustrates the timing constraints. The maximum sampling frequency now can be calculated by

$$F_s = \frac{1}{4(N \text{ channels})(T_1 + T_2 + T_3)}$$

Note: A factor of 4 is used for margin to account for parasitics and other circuit imperfections.

For maximum system performance, times T_1 and T_2 can be significantly reduced by employing two sample-and-holds. While one S/H is in the hold mode and a conversion is under way, the other S/H can acquire another input channel. This can be accomplished by sharing the input buffer amplifier and using a switch arrangement to switch in one of two output holding amplifiers.

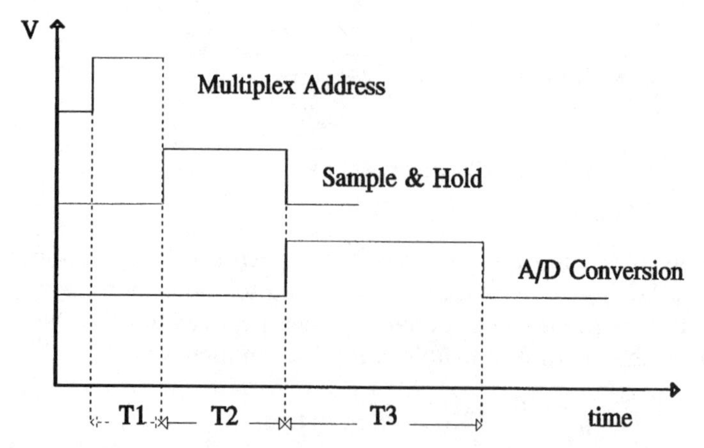

Figure 3.16 System timing constraints where T_1 = multiplexer address, transition, and settling time; T_2 = S/H acquisition time, including hold-mode settling time; and T_3 = A/D conversion time including the time to access the data.

Analog Switch Applications

Analog switches, as mentioned earlier, are very similar in nature to a multiplexer; however, they can serve different purposes. One of the most common uses of an analog switch is to alter the gain of an amplifier. When a switch is used in this fashion, it is important to eliminate the possibility of the on resistance affecting the gain (as in Fig. 3.17). This problem can be completely eliminated by placing the switch within the high-impedance loop of the amplifier. An example of this was shown previously in the two op amp instrumentation amplifier circuit of Fig. 3.5.

Another common use of analog switches involves controlling different input signals to an amplifier (i.e., an integrating amplifier used for a dual-slope A/D converter). In Fig. 3.18, the unknown input sig-

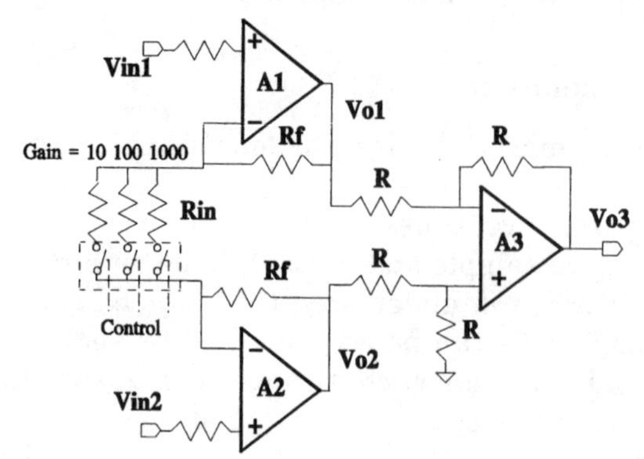

Figure 3.17 Analog switch is used to set gain control.

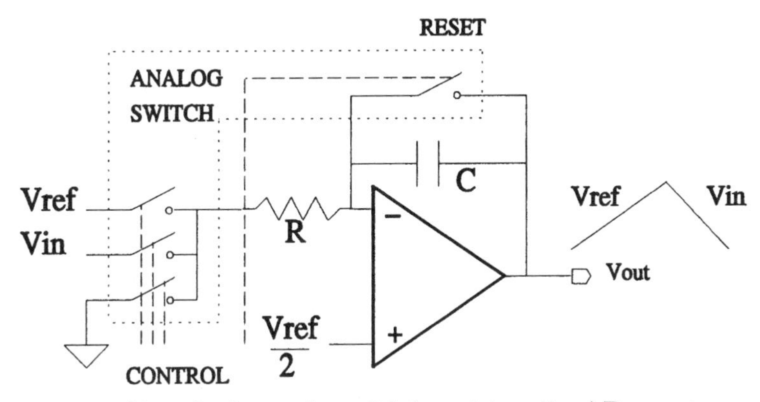

Figure 3.18 Example of an analog switch for an integrating A/D converter.

nal is first integrated for a fixed number of counts (resolution); then a known reference of opposite polarity is switched in to force the integrating ramp to approach ground. Besides switching inputs, the analog switch can be used to reset the integrating capacitor prior to a conversion cycle by providing a short circuit across the capacitor.

Filters

No chapter on active support components would be complete without a discussion of filters. Filters are required prior to the A/D converter to attenuate not only high-frequency input signals but also system noise. The primary function is to avoid aliasing (discussed in Chap. 1), where the input signal bandwidth must be limited to below f_s /2 (Nyquist frequency F_n). In practical systems this really needs to be at least F_s /8 depending on the effectiveness of the antialiasing filter. There are numerous options in constructing a filter. In each case, the application and type of analog-to-digital converter (ADC) used will make a significant impact on the filter design. Before you set out to design a filter, you should review several key filter characteristics and system concerns:

Key filter characteristics:

1. High attenuation at the Nyquist frequency
2. Low noise
3. Flat gain within the desired bandwidth
4. Low distortion
5. Adequate step response
6. Low phase error

System concerns:

1. Signal noise spectrum
2. Desired measurement bandwidth
3. Type of ADC chosen

Integrating ADCs, by their nature, can easily deal with noise or higher-frequency components along with the desired input signal. This is due to the fact that the input is applied during the entire integration cycle. In other words, the input signal is not sampled just prior to a conversion; it is constantly being measured. Other converters (i.e., flash and successive-approximation), on the other hand, place much greater demand on the antialiasing filter since these converters sample or convert within a short time. The faster the ADC, the easier the filter design, however. This is due to the fact that the higher cutoff frequency means smaller filter components. When a multiplexer is used to switch input signals with large differences in amplitude, it is important to lessen the phase delay or settling time between channels. This usually means that the Bessel-type low-pass filter makes a good choice.

There are many excellent books on filtering techniques. Filters can consist of a simple resistor-capacitor low-pass arrangement, active filter with an op amp, or a switch capacitor filter. In addition to providing a low-pass filter, the amplifier can be used to gain the signal, as is the case with an active filter.

As discussed in Chap. 8, one of the major benefits of the delta-sigma A/D converters is the relaxed filter requirements. This is due to the ease of implementing a high-order digital filter to prevent aliasing or system noise. It is very common for delta-sigma converters to be used with either a simple *RC* filter or, in some cases, no external filter at all.

Systems that include other constant operating frequencies may require a low-pass filter along with a notch (or bandstop) filter. For instance, this could attenuate the line frequency (that is, 60 or 50 Hz), processor oscillator, or switching regulator interference from the signal of interest. Systems that include a digital signal processor can use a digital notch filter to eliminate the line frequencies or any other constant frequency within the desired bandwidth.

Chapter

4

Low-Cost A/D Conversion Techniques With Microcontrollers

Many products currently take advantage of a microcontroller and are capable of creating an analog-to-digital (A/D) converter with very little extra hardware or software. In an effort to make the design more competitive, the typical successive-approximation A/D converter (ADC) on board can be replaced by several low-cost options. The designer must first determine the true system requirements and then select an approach that best meets those requirements.

The most common misconception involves the speed required for A/D conversion. Rarely is it necessary for the actual conversion to be completed within the time frame of typical on-board A/D converters. What is usually far more important than the A/D conversion speed is that the total workload of the microcontroller is not appreciably affected. Consider the fact that the microcontroller usually has several thousand ROM bytes to loop through between A/D converter measurements. Then why should the A/D converter measurement have to be completed within a few microseconds?

If speed is really important due to the complete system response time, then there are still options such as choosing a relatively software-independent approach. This, of course, does not mean that a discrete approach is going to fit every application. In some cases, the signal to be measured can be relatively fast-changing in comparison to the ADC speed. Although an external sample-and-hold can solve this problem, this probably means that a discrete ADC is not cost-effective. However, even in these situations there are some software tricks that can be used. For example, the motor current that is being controlled by a 20-kHz pulse-width modulation (PWM) can be measured by extending the drive momentarily once every software loop. This allows the ADC measurement to be completed over a longer pulse time and still not affect the motor current appreciably, since the net motor duty cycle is not impacted. In other words, the motor current probably does not have to be measured every pulse period.

Quite often when a discrete ADC is designed, tradeoffs must be made between parameters such as speed versus resolution and high accuracy versus low cost. Besides the actual signal parameters (i.e., rate of change, accuracy, resolution, and range), other tasks that need to be performed by the microcontroller often play a major role in which method is optimum. System performance requirements such as real-time monitoring/control may dictate which approach to use based on system response times. For example, it is important to understand how fast the input sensors can change and how fast the output devices can (or need to) react. Other factors to consider are the amount of software and external hardware required for each approach. After all these factors have been taken into account, an optimum solution can often be reached resulting in considerable cost savings compared to a fully integrated solution.

The following sections describe several A/D conversion techniques to choose from. Although the examples represent the most common configurations, undoubtedly many variations can be made to better fit specific requirements. With some engineering creativity, the examples described in this chapter may inspire some new approaches. For a solution to be considered practical, it must be able to meet several criteria.

Criteria for discrete ADC design:

1. Costs less than fully integrated device
2. Meets all necessary requirements (i.e., resolution, accuracy, and speed)
3. Uses minimal external hardware or software
4. Utilizes available I/O
5. Meets long-term drift error requirements
6. Does not require calibration (except when sensor tolerance calls for it)

Although applications that can tolerate a slower A/D conversion process are likely candidates for a discrete design, they should not be restricted to slow systems. As you will see in the following sections, some of the techniques can be quite fast, if necessary. Below are just some of the typical applications that can utilize a lower-cost discrete approach.

Consumer products (temperature control for irons, ovens, thermostats, coffee makers, etc.)

Automotive products (climate control for temperature; pressure sensors; operator controls for fans, wipers, radio, etc.; photosensors for automatic-lamp, general-purpose diagnostics, and compass sensor for navigation)

Commercial products (process control of temperature and pressure, laptop computer battery monitor)

Conventional Pulse-Width Modulation A/D Conversion

Many engineers are familiar with the conventional approach to performing the pulse-width modulation (PWM) A/D conversion. This involves using a fixed-frequency square-wave signal with a variable duty cycle to average out through a low-pass filter. The average voltage is then compared to an unknown voltage via a comparator. Providing the low-pass filter (typically an *RC* network) has a time constant much longer than the PWM period, the capacitor ripple voltage will be relatively small. This will produce minimal error and provide direct correlation between the duty cycle and the unknown input voltage. Assuming that the PWM signal goes from 0 to V_{ref}, the input can be determined from the following:

$$V_{in} = V_{ref} \times \text{duty cycle}$$

Given the PWM period T (as the time base), the desired resolution, and voltage range V_{ref}, the resistor and capacitor can be chosen from the basic equation

$$V_c(t) = V_{ref}\left(e^{-T/RC}\right)$$

The period T is constrained by the microcontroller clock frequency and the desired resolution. For example, with a 1-μs instruction cycle time and 8 bits of resolution, T will equal 256 μs. Since it is desired to have no more than 1 least significant bit (LSB) of ripple voltage, the V_c / V_{ref} ratio for an 8-bit measurement equals $^{255}/_{256}$. Now, solving the above equation for RC by taking the natural logarithm of both sides yields

$$RC = \frac{-T}{\ln[(n - 1)/n]}$$

where n = resolution. For an 8-bit measurement, RC equals 256 μs divided by $\ln {}^{255}\!/_{256}$, or − 0.00391 for a 65-ms RC time constant! This is a worst-case situation since it is assumed that the duty cycle is very close to zero, where the decay time between pulses is the longest. Depending on the desired accuracy and PWM step changes, several time constants probably need to pass before new duty cycles can be tried.

Use of this technique requires that the input signal being measured change very slowly. This could be the case when temperature or battery voltage is measured, for example. Sufficient time must be allowed for the capacitor to either charge or discharge to within 1 LSB of the final voltage before the duty cycle is readjusted. If large step changes occur in the input signal, the conversion time may take up to about 500 ms or more.

The example circuit in Fig. 4.1 uses a software-independent PWM timer along with an on-board comparator. This has the advantage of allowing the microcontroller to check the comparator at convenient times while performing other tasks. Each time the comparator is checked, the duty cycle can be slowly changed in the direction to make the average capacitor voltage equal the input.

Possible error sources would include the comparator offset voltage V_{os} and the microcontroller voltage drops when switching to ground, or V_{supply}. The error from the comparator is simply V_{os}. The error caused by the microcontroller voltage drops is not as simple to calculate. Fortunately, this effect is reduced by the actual voltage drop times the percentage of time that the microcontroller is at that level. What this means is that the error will tend to cancel at midrange and will be small near the low or high range. For example, with inputs near ground, the capacitor voltage would have a voltage very close to

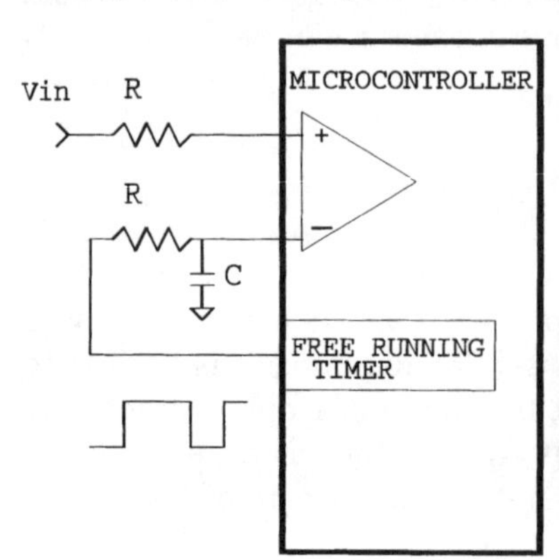

Figure 4.1 Basic PWM A/D converter with software-independent timer and comparator.

ground (equal to V_{in} ± 1 LSB). Therefore, the average load on the microcontroller output will be very light for the majority of the pulses since most would have to also be at ground. If the microcontroller of choice does not meet the desired accuracy, an external comparator and/or a CMOS buffer for driving the *RC* network can be used.

The following section describes a greatly improved PWM approach that utilizes the same hardware. In most cases, the improved PWM A/D converter will make a better choice. There is only one exception where the conventional approach may be preferred. This occurs when the signal to be measured is nearly a dc level and the microcontroller can very infrequently monitor the signal and vary the PWM drive. Note that this would provide a very light workload on the microcontroller allowing more time for all the other tasks at hand, providing a software-independent PWM timer is used. In other words, the A/D converter can be read almost instantly each software loop.

Improved PWM A/D Conversion

Previously described, the "conventional" PWM technique can take more time than desired. Modifying the technique and increasing the PWM frequency will significantly speed up the conversion. Instead of relying on a single, long period with a duty cycle, summing several shorter pulses allows a filter with a much smaller *RC* time constant to be used, thus eliminating the long settling delays associated with the previous PWM method. This new technique works by averaging (or summing) several short pulses of equal duration over a fixed time. In Chap. 8 it will become apparent that there are a lot of similarities between the improved PWM A/D conversion method and the delta-sigma A/D conversion method. The following equations describe basically how the improved PWM A/D converter works.

$$V_{in} = V_{ref}\left(\frac{\text{sum of high pulses}}{\text{sum of high and low pulses}}\right)$$

$$\text{Conversion time} = \text{resolution} \times \text{single pulse duration}$$

It is important to understand that for a linear measurement the high and low pulses need to have the exact same duration. Later discussions show how deliberately extending either high or low pulses can provide nonlinear curve matching. Fortunately, the accuracies of the high and low pulse durations are affected not by the type of oscillator (that is, *RC* or crystal) but by the number of instruction cycles within the software loops. When the pulse durations are the same, an equal weight charge is either added for a high pulse or subtracted for a low pulse, based on the difference between the capacitor and the pulse voltage levels. Actual conversion times will depend on the microcontroller oscillator clock (affects pulse duration) and the desired resolution. For example, with a 20-μs software loop for high and low pulse durations and an 8-bit resolution, the conversion time will be about 5 ms (255 × 20 μs). In addition, the time to initialize the capacitor must be accounted for. Depending on how you choose to do this, it can take an additional 100 μs to 5 ms.

An intuitive explanation of the improved PWM A/D conversion technique can be given by using some simple concepts. Assume that the sampling of the comparator is fast enough compared to the *RC* time constant. Then the voltage on the capacitor can be made to track the input signal within a closed-loop system made up of the comparator, microcontroller, and *RC* network. Now if we start the conversion process with the capacitor initialized close to V_{in}, the sum of high and low pulses that are subsequently applied must indicate the unknown V_{in} after some elapsed period. This is because of the fact that we are maintaining the voltage on the capacitor, and the only way to do that is to provide a net duty cycle that averages out to V_{in}. The following mathematical discussion is for those analytical types and further describes how it works.

Mathematical analysis. Let

$$n = \text{total number of high pulses } (T_{on})$$
$$m = \text{total number of low pulses } (T_{off})$$
$$V_c = \text{capacitor voltage}$$
$$V_{hi} = \text{high portion of output voltage}$$
$$V_{lo} = \text{low portion of output voltage}$$

then

$$V_c(t) = V_c + n(V_{hi} - V_c)\left(1 - e^{-t/RC}\right) - m(V_c - V_{lo})\left(1 - e^{-t/RC}\right)$$

Let

$$V_c = V_{in} \text{ at start of conversion}$$
$$V_c(t) = V_{in} \text{ within a tight tolerance}$$
$$K = 1 - e^{-t/RC}$$

then

$$V_{in} = V_{in} + KnV_{hi} - KnV_{in} - KmV_{in} + KmV_{lo}$$

$$0 = KnV_{hi} - KnV_{in} - KmV_{in} + KmV_{lo}$$

Let

$$V_{hi} = V_{ref} - V_{os}$$

where V_{os} = voltage drop from voltage supply to output. Then

$$0 = KnV_{ref} - KnV_{os} - KnV_{in} - KmV_{in} + KmV_{lo}$$

Arranging like terms and solving for V_{in} give

$$V_{in}(n + m) = V_{ref}(n) - nV_{os} + mV_{lo}$$

$$V_{in} = V_{ref}\left(\frac{n}{n + m}\right) - \frac{nV_{os} - mV_{o}}{n + m}$$

Note: The conversion actually starts after the capacitor voltage V_c is equal to V_{in}. There are several ways to initialize the capacitor, which are discussed later. Also note that the K term $[1 - e^{-T/RC}]$ drops out of the equation, and therefore the tolerances of the resistor and capacitor will not affect accuracy.

In the above equation for V_{in}, the left portion represents the nonerror term, and the right portion is the potential error. This error, however, is relatively small when a CMOS microcontroller is used to drive the *RC* network. CMOS microcontrollers typically switch very close to the power and ground levels with a light load (approximately 50 µA). In addition, there is a cancellation effect near the midrange where n approximately equals m. When the duty cycle is at a low or high percentage, the microcontroller output load is minimized due to a very small voltage difference between the capacitor and the average voltage from the microcontroller output pin. The effective offset voltages are further reduced by the percentage of time that the pulse is present. For example, given V_{os} = 15 mV from the positive supply, V_{lo} = 10 mV from ground, and a 20 percent duty cycle ($n = 0.2$, $m = 0.8$), the effective error is

$$V_{error} = 0.8(10 \text{ mV}) - 0.2(15 \text{ mV}) = (8 - 3) \text{ mV} = 5 \text{ mV}$$

For an 8-bit conversion of a 5-V input range, this would represent only ¼ LSB!

Figure 4.2 illustrates the relationship between the microcontroller square-wave PWM output and the ripple voltage on the capacitor. Note that the capacitor voltage V_c is centered on the input voltage V_{in} and that the triangular waveform is exaggerated for illustrative pur-

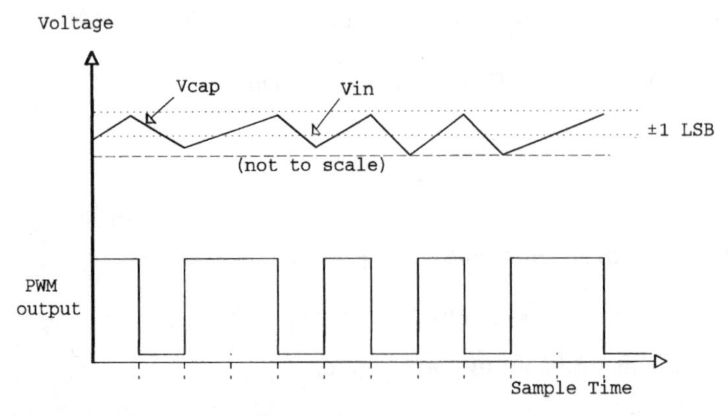

Figure 4.2 Improved PWM and V_c waveforms.

poses. At regular intervals, the comparator is sampled to determine the level to drive the *RC* network for equalizing V_c and V_{in}. Since the pulse durations are much shorter than the *RC* time constant, the waveforms will be very linear.

Hardware concerns

This improved PWM technique replaces the free-running hardware timer in the conventional PWM approach with a software timer. Figure 4.3 illustrates the basic circuit for creating the PWM A/D converter. Depending on the comparator input bias (or leakage) current, a tradeoff may have to be made in choosing the external resistor. This tradeoff is between using a large resistance to minimize the load on the microcontroller output and a small resistance to reduce the comparator offset voltages. When the on-board CMOS comparator is used, the input leakage current over temperature should be checked to ensure that the chosen resistance does not cause undesired error.

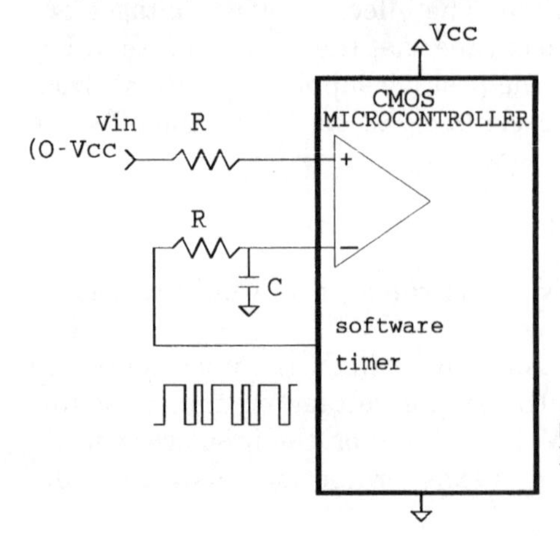

Figure 4.3 The COP820CJ basic PWM circuit.

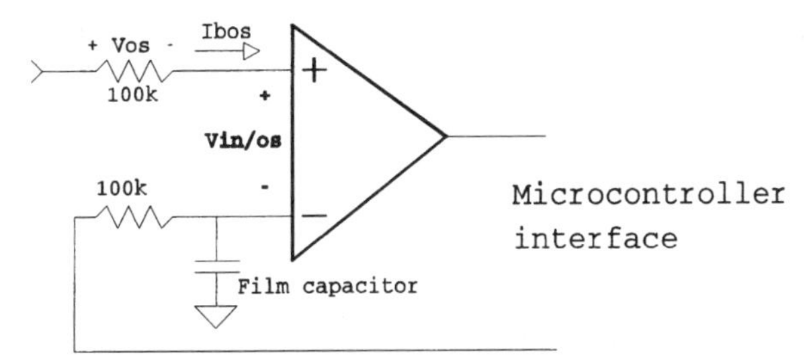

Figure 4.4 Bipolar comparator with 100-kΩ resistors.

Regardless of whether an internal or external comparator is used, the input bias or leakage current may require that the value of the resistance be lower. The important parameter for determining the error from this input resistance is the input bias offset current (I_{bos}). Since a resistance of equal value can be chosen for both inputs, only the difference in input bias currents will cause error. For example, in Fig. 4.4, where a bipolar comparator is used with I_{bos} = 200 nA over temperature and 100-kΩ resistors for both inputs, the offset voltage equals

$$V_{\rm os} = RI_{\rm bos} = 100\ \text{k}\Omega(200\ \text{nA}) = 20\ \text{mV}$$

Keep in mind that comparators have a limited input voltage range. With a 5-V supply, the COP820CJ comparator can work from ground to about 4 V. When the common LM339 comparator is used, the range is ground to V_{cc} − 1.5 V, or 3.5 V.

Once the resistor is selected, the capacitor should be chosen so that the ripple voltage on the capacitor is about 1 LSB during a single pulse period. The actual capacitance value (and tolerance) is not critical as long as the ripple voltage is in the range of about ½ to 2 LSB. Here is an example.

Determine the capacitor value:

- Single-pulse durations = 20 µs.
- Range = 0 to 5 V.
- Select R = 100 kΩ.
- Accuracy = 8 bits (1 LSB = $^5/_{256}$ = 19.5 mV).

From $i = \dfrac{C\,dv}{dt}$:

$$\text{C} = \frac{5\ \text{V}}{100\ \text{k}\Omega(20\ \mu\text{s}\ /19.5\ \text{mV})} = 0.051\ \mu\text{F}$$

Select C = 0.047 µF. *Note:* You can assume a linear slope during a single sample time.

What is important about the capacitor is the leakage resistance. It is recommended that a film capacitor be used with high insulation resistance (i.e., compared to a ceramic disk). Refer to Chap. 2 concerning passive support circuits.

Basic PWM A/D Conversion Software

As illustrated in the flowchart and code listing in Figs. 4.5 and 4.6, respectively, the conversion process is started by initializing the TOTAL and TON software counters for maximum counts. The TOTAL counter keeps track of the number of times the comparator is sampled, and the TON counts the number of times that the *RC* network is hit with a high pulse. At equal intervals, the comparator sample will determine the polarity of pulse required to equalize the capac-

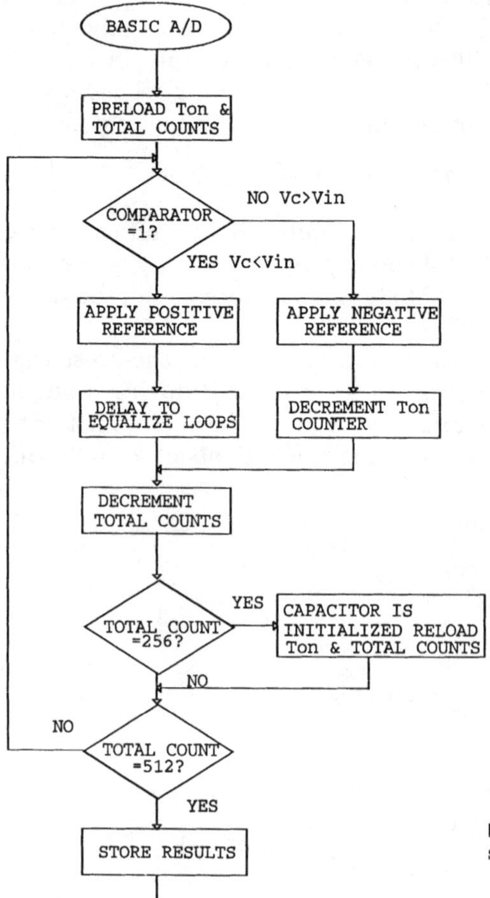

Figure 4.5 Basic PWM A/D conversion flowchart.

```
;The program listed below will work in any COP800 microcontroller
;(i.e. COP820, COP840, COP880, COP888). SET UP FOR .047mfd CAP.,
;100K RES, @1 MICRO. CYCLE TIME. THE FIRST CONVERSION
;L0=COMPARATOR OUTPUT, L1=PWM DRIVE, Vin=(+) COMP. INPUT.
;INITIALIZES, AND 2nd IS THE RESULT STORED IN RAM LOCATION 00.
.CHIP 820
LCONF=0D1
LDATA=0D0
TON=0F2
TOTAL=0F0
PSW=0EF
;
        LD A,#02      ;USE TO DETERMINE WHEN TO RELOAD
        LD TOTAL,#0FF ;PRELOAD TOTAL COUNTS
        LD 0F1,#2  ;MULTIPLIER (255 TO INIT. PLUS 255 FOR RESULT)
        LD TON,#0FF ;PRELOAD Ton
        LD 0FE,#0D0 ;LOAD B REG TO POINT TO LDATA REG.
        LD LDATA,#01  ;L PORT DATA REG, L0=WEAK PULL UP, L1=HIGH
        LD LCONF,#02  ;L PORT CONFIG REG, L0=INPUT, L1=OUTPUT
LOOP:   IFBIT 0,0D2 ;TEST COMPARATOR OUTPUT
        JP HIGH     ;JUMP IF L0=1
        NOP
        NOP         ;EQUALIZE TIME FOR SETTING AND RESETTING
        RBIT 1,[B]  ;DRIVE L1 LOW
        DRSZ Ton    ;DECREMENT Ton WHEN DRIVING LOW
        JMP COUNT
HIGH:   SBIT 1,[B]  ;DRIVE L1 HIGH
        NOP
        NOP
        NOP
        NOP
        NOP
        NOP         ;EQUALIZE HIGH AND LOW LOOPS
COUNT:  DRSZ TOTAL  ;DECREMENT TOTAL COUNTS
        JP LOOP
        RBIT 1,LCONF ;TRISTATE L1 TO MINIMIZE ERRORS FROM EXTRA
        RBIT 1,[B]    ;CYCLES
        IFEQ A,0F1  ;CHECK INITIALIZATION LOOP COMPLETE
        JP RELOAD    ;JUMP IF TRUE.
        JP DEC       ;JUMP IF NOT END OF 2nd LOOP
RELOAD: LD 0F2,#0FF ;RELOAD Ton WITH FF
        LD 0F0,#0FF ;SYNC TOTAL AND Ton COUNTERS
DEC:    SBIT 1,[B]  ;SET L1 HIGH
        SBIT 1,LCONF  ;RESTORE L1 AS OUTPUT.
        DRSZ 0F1    ;DECREMENT MULTIPLIER UNTIL ZERO
        JMP LOOP  ;CONTINUE A/D UNTIL AFTER 2nd CONVERSION
        LD A,TON  ;LOAD A WITH Ton
        X A,00    ;STORE RESULT IN RAM LOCATION 00
.END
```

Figure 4.6 Code listing.

itor voltage and the input voltage. By taking advantage of a multipurpose instruction (DRSZ), the TON counter is decremented each time a negative (or low) pulse is produced. Every sample time, the TOTAL counter is also decremented.

Note the NOP delays in the positive and negative pulse software loops. This is due to the requirement that each pulse be the exact same number of cycles (unless nonlinear measurements are desired). Some extra care needs to be taken here to make sure that regardless of the previous state of the output pulse, the next pulse will have the proper duration. The number of instruction cycles from the comparator check (IFBIT 0,0D2) to the instruction where the output is either reset (RBIT 1,[B]) or set (SBIT 1,[B]) has to be the same (in this example it is 8 cycles).

When the TOTAL counter is equal to 256, the capacitor will be initialized close to V_{in}. Remember from the mathematical analysis that $V_c = V_{in}$ before the conversion. What we are really doing is merely keeping track of the number of times we need to add or subtract charge from the capacitor to maintain $V_c = V_{in}$ over a fixed period.

Options for initializing the capacitor

Prior to the actual conversion, the capacitor can be initialized in a number of ways. The initial 256 counts in this example to initialize the capacitor are adequate due to the tristate of the *RC* drive output pin on the microcontroller after each conversion. You can think of this as completing a dummy conversion and then reloading TON and TOTAL counters for the actual measurement. In the multichannel discussion, there is an example of a capacitor speed-up technique using two diodes that significantly speeds up the initialization process. With the diodes, only a small fraction (about 50 pulse durations) will be required to force the capacitor close to V_{in}.

Another approach uses an analog switch (i.e., the HC4066). As shown in Fig. 4.7, this could be used to switch the input voltage to the capacitor,

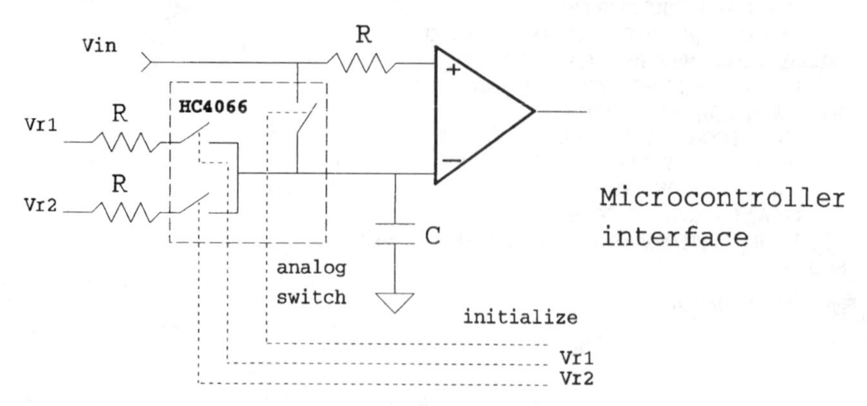

Figure 4.7 Analog switch for capacitor initialization and reference.

resulting in a very fast initialization time. Note that the analog switches can also be used for switching between the low and high references. This provides a slightly more accurate conversion due to lower voltage offsets when driving the *RC* network. Another advantage is the flexibility of using references that are different from the microcontroller V_{cc} and ground. This provides a hardware version of the variable-range PWM A/D converter that is discussed later in this chapter. By using either a hardware or a software variable-range technique, conversions can be easily restricted to the desired limits of the input signal, thus utilizing the maximum resolution available and providing maximum speed.

Multichannel PWM A/D Conversion

When multiple analog signals need to be measured, a different approach may be required for initializing the capacitor. One approach is to use a multiplexer along with two diodes connected from the output of the multiplexer to the capacitor. The optional diodes are for speeding up the capacitor initialization. Each channel could have different voltages; thus tristating between measurements would not maintain the voltage for the next channel. In Fig. 4.8, the HC4051 8:1 multiplexer has 2-kΩ input resistors to minimize the current into the MUX and provide a quick charge path through the diodes. The diode will continue to conduct until the voltage drop across it is about 0.35 V, at which time only microamperes will be flowing. The diode leakage current will not be a factor since the technique requires that the capacitor voltage stay near V_{in} to within about 1 LSB. In other words, the voltage drop across the diodes will be near 0 V. Once the diode current tapers off, the PWM signal from the microcontroller will complete the initialization process well before the 256 pulses used in this example.

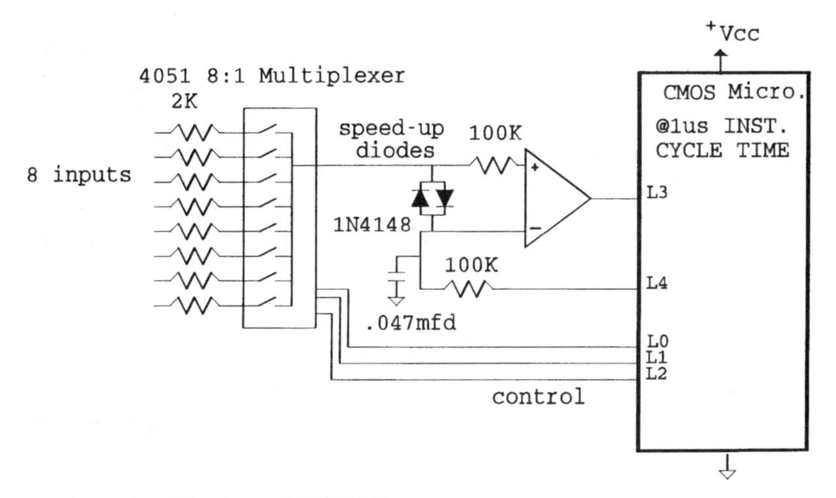

Figure 4.8 An eight-channel PWM A/D converter.

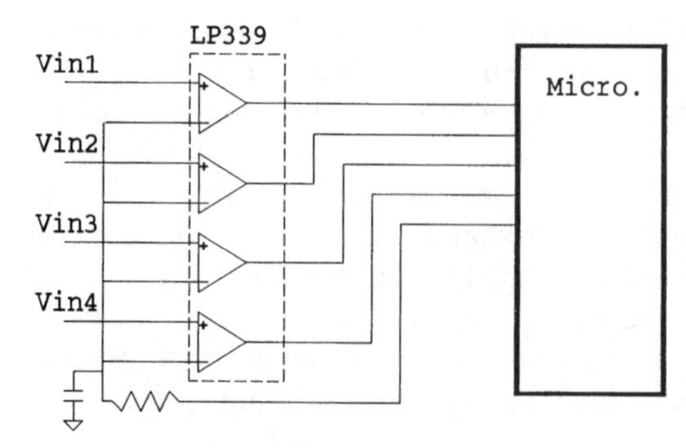

Figure 4.9 Four-channel quad comparator.

Another approach to measuring multiple channels is to feed all four comparators of a quad package like the LP339 with the same *RC* network (see Fig. 4.9). Only one channel can be measured at a time with this approach, and the speed-up diodes cannot be used. Thus, extra time will be required to initialize the capacitor prior to each conversion.

An advantage of the LP339 quad comparator is that the offset bias current I_{bos} is extremely low (5 nA). This will not cause any noticeable voltage drop (error) from the four comparator input bias currents through the common resistor. It makes sense to use the multiplexer if maximum speed is necessary or if more than four channels must be measured.

Multichannel PWM A/D Conversion Software

Software for measuring up to eight channels by using a CD/HC4051 is very similar to that for a single channel. Referring to the flowchart in Fig. 4.10 and the code listing in Fig. 4.11, we see that the differences are in the beginning and end. As can be seen in the code listing, there are minimal additional instructions for dealing with the multiplexed channels and corresponding RAM storage locations. Before the conversion is begun, the X pointer register is set to RAM location 00 and the channel is set to channel 0. Each time a conversion is completed on an input channel, the storage RAM location and the input channel are incremented. The RAM pointer and associated channel select are then reset when the eighth channel is completed.

Variable-Range PWM A/D Conversion*

In previous discussions of the PWM A/D conversion technique, pulses were either at ground or at V_{supply} for the entire pulse duration. By

*U.S. patent 5,189,421.

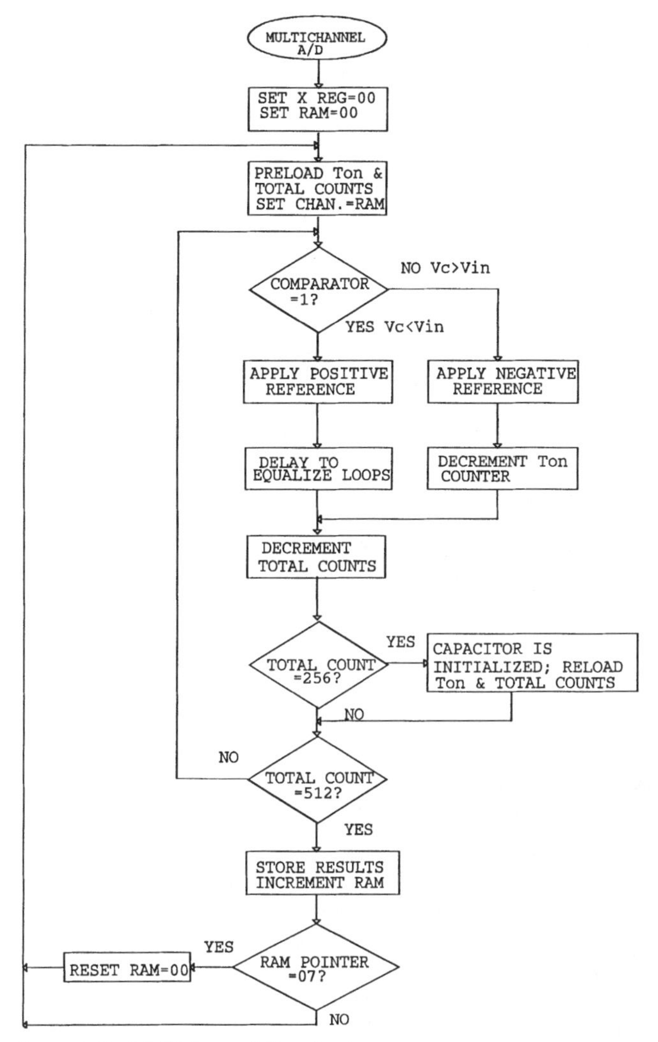

Figure 4.10 Multichannel flowchart.

changing the individual pulses from a static level to a PWM signal, the range of the conversion can be modified totally in software. You can think of this as a PWM signal within a PWM signal of lower frequency. The basic principle of operation is exactly the same as before. For example, say that a signal to be measured to within 1 percent is in the range of 1.10 to 3.20 V. With a 5-V supply for V_{ref} this signal range is accommodated by defining a low pulse equal to 5 cycles high and 19 cycles low and a high pulse equal to 16 cycles high and 8 cycles low, as shown in Fig. 4.12. Note that both the high and low pulse periods are set to the same 24 counts. This is important if a linear response is desired.

```
;THIS IS AN 8 BIT, 8 CHANNEL PWM A/D PROGRAM, AND USES A
;CD4051 WITH A COMPARATOR. RESULTS ARE STORED IN RAM 00-07.
;THE 1st CONVERSION INITIALIZES, AND THE 2nd IS THE RESULT.
.CHIP 820
LDATA=0D0
LCONF=0D1
TON=0F2
TOTAL=0F0
;
        LD X,#00    ;INITIALIZE X REG FOR 1st RAM LOC.
CONVER:  LD TOTAL,#0FF ;PRELOAD TOTAL COUNTS
        LD 0F1,#02  ;TOTAL LOOP COUNTER
        LD TON,#0FF ;PRELOAD TON
        LD 0FE,#0D0 ;INIT. B REG TO POINT TO Ldata REG
        LD LDATA,#018  ;Ldata, L0-2=LOW, L3=PULLUP, L4=HIGH
        LD A,X      ;USE CURRENT RAM POINTER TO SELECT -
        OR A,LDATA  ;PROPER A/D CHANNEL.
        X  A,LDATA  ;MODIFY Ldata FOR CHANNEL SELECTION.
        LD LCONF,#017  ;Lconf. REG. L0-L2,L4=OUTPUT,L3=IN
LOOP:    IFBIT 3,0D2 ;TEST COMPARATOR OUTPUT AT L3 INPUT
        JMP HIGH    ;JUMP IF L3=HIGH
        NOP
        NOP         ;EQUALIZE TIME FOR SET AND RESET
        RBIT 4,[B]  ;DRIVE L4 LOW WHEN COMPARATOR IS LOW.
        DRSZ TON    ;DECREMENT Ton WHEN APPLYING NEG. REF.
        JMP COUNT   ;JUMP TO COUNT UNLESS TON REACHES ZERO
HIGH:    SBIT 4,[B]  ;DRIVE L4 HIGH WHEN COMPARATOR IS HIGH
        NOP
        NOP
        NOP
        NOP
        NOP
        NOP         ;EQUALIZE HIGH AND LOW LOOP TIMES
COUNT:   DRSZ TOTAL  ;DEC. TOTAL COUNTS EACH LOOP
        JMP LOOP    ;JUMP UNLESS TOTAL CNTS.=0
        RBIT 4,LCONF ;TRISTATE L4 TO MINIMIZE ERROR
        RBIT 4,[B]  ; "              "
        LD A,#02    ;USE TO DETERMINE WHEN TO RELOAD
        IFEQ A,0F1  ;CHECK FOR 2nd CONVERSION COMPLETE
        JP RELOAD   ;IF TRUE
        JP DEC      ;OTHERWISE JUMP TO DEC
RELOAD:  LD TON,#0FF  ;RELOAD TON FOR START OF NEXT CONV.
        LD TOTAL,#0FF ;SYNC TON AND TOTAL COUNTS
DEC:     SBIT 4,[B]  ;SET L4 HIGH
        SBIT 4,LCONF ;RESTORE L4 AS OUTPUT
        DRSZ 0F1    ;DECREMENT TOTAL LOOP UNTIL ZERO.
        JMP LOOP    ;DONE WHEN 0F1 IS ZERO.
        LD A,TON    ;LOAD A WITH Ton RESULT
        X A,[X+]    ;STORE RESULT AT CURRENT RAM POINTER
                    ;AND AUTO INCREMENT POINTER
        LD A,#08    ;CHECK [X] RAM POINTER FOR
        IFEQ A,X    ;EIGHTH CHANNEL CONVERTED
        LD X,#00    ;RESET RAM POINTER IF [X]=8
        JMP CONVER
.END
```

Figure 4.11 Multichannel code listing.

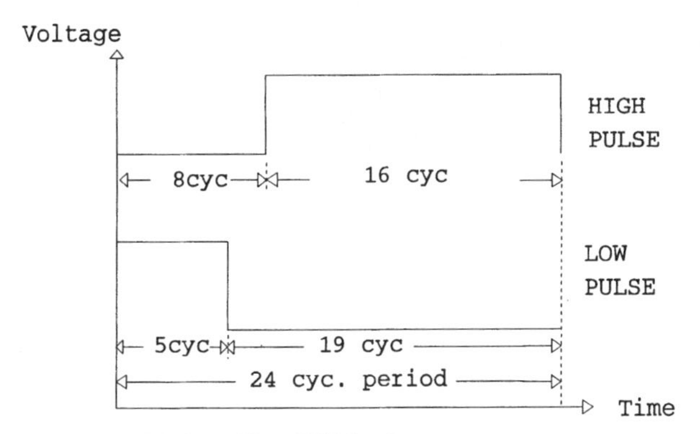

Figure 4.12 High and low PWM pulses.

These pulses provide a lower range equal to zero counts of (5 V)(5⁄24), or 1.042 V, and an upper range equal to 100 counts of (5 V)(16⁄24), or 3.333 V. Usually, all that is needed is the counts, so no further computations are normally required in order to use the result. If the actual voltage is desired (i.e., for display), it can be computed by

$$V_{\text{in}} = V_L + (V_H - V_L)(T_{\text{on}} \text{ counts/total counts})$$

where V_L = average low-pulse voltage
V_H = average high-pulse voltage

Many modifications of the above example are possible. Changing the duty cycles of the pulses will alter the range. Changing the total counts will provide a different amount of resolution. There are two major benefits to this approach. First, the speed of the measurements can be increased owing to the ability to measure only the range of operation. (What is the point of measuring a car battery down to 0 V?) Second, the resolution can be significantly increased. A rough estimate of the unknown signal can be used to alter the duty cycles for the low and high pulses. For example, a 5-V system can be broken up into four ranges, using only 8 bits for a 10-bit measurement. If the rough estimate indicates that the signal is about 2 V, then the software can apply a series of pulses with low = 6 high and 18 low cycles and high = 12 high and 12 low cycles. This provides a conversion range of 1.250 to 2.500 V for an effective 10-bit resolution using only 8 bits in a limited range.

The software for this example is shown in Fig. 4.13, and the flowchart is exactly the same as before, with the only exception in the software being the duty cycles in the drive pulses.

```
;THIS IS AN 8 BIT VARIABLE RANGE PWM A/D PROGRAM.
;THIS ROUTINE PROVIDES THE FULL 8 BITS WITH A SINGLE 5 VOLT
;SUPPLY FOR THE COMPARATOR AND MICRO.  NOTE THAT THE
;COMPARATOR TYPICALLY HAS A COMMON MODE INPUT RANGE OF 0 TO
;(Vcc-2V), OR (0 TO 3) VOLTS WITH A 5V SUPPLY.  THE POSITIVE
;PULSE=9 CYCLES LOW & 14 HIGH, NEG.  = 23 LOW.
.CHIP 820
LDATA=0D0
LIN=0D2
LCONF=0D1
TON=0F2
TOTAL=0F0
;
CONV:   LD A,#02    ;USE FOR COUNTING TOTAL LOOPS
        LD 0F1,#02  ;TOTAL LOOP COUNTER
        LD TOTAL,#0FF ;PRELOAD TOTAL
        LD TON,#0FF ;PRELOAD TON
        LD 0FE,#0D0 ;INIT. B REG TO POINT TO Ldata REG
        LD LDATA,#01  ;Ldata, L0=PULLUP, L1=HIGH
        LD LCONF,#02  ;Lconf. REG. L0=INPUT, L1=OUTPUT
LOOP:   IFBIT 0,LIN ;TEST COMPARATOR OUTPUT AT L0 INPUT
        JP HIGH     ;JUMP IF L0=HIGH
        NOP
        NOP         ;EQUALIZE TIME FOR SET AND RESET
        RBIT 4,[B]  ;DRIVE L4 LOW WHEN COMPARATOR IS LOW.
        DRSZ TON    ;DECREMENT Ton WHEN APPLYING NEG. REF.
        NOP
        NOP
        NOP
        JMP COUNT   ;JUMP TO COUNT UNLESS TON REACHES ZERO
HIGH:   RBIT 1,[B]  ;RESET L1 FOR TOTAL OF 9 CYCLES
        NOP
        NOP
        NOP
        NOP
        NOP
        NOP
        NOP
        NOP
        SBIT 1,[B]  ;DRIVE L1 HIGH FOR TOTAL OF 14 CYCLES
COUNT:  DRSZ TOTAL  ;DEC. TOTAL COUNTS EACH LOOP
        JMP LOOP    ;JUMP UNLESS TOTAL CNTS.=0
        RBIT 1,LCONF ;TRISTATE L4 TO MINIMIZE ERROR
        RBIT 1,[B]  ; "              "
        IFEQ A,0F1  ;CHECK FOR 2nd CONVERSION COMPLETE
        JP RELOAD   ;IF TRUE
        JP DEC      ;OTHERWISE JUMP TO DEC
RELOAD: LD TON,#0FF ;RELOAD TON FOR START OF NEXT CONV.
        LD TOTAL,#0FF ;SYNC TON AND TOTAL COUNTS
DEC:    SBIT 1,[B]  ;SET L4 HIGH
        SBIT 1,LCONF ;RESTORE L4 AS OUTPUT
        DRSZ 0F1    ;DECREMENT TOTAL LOOP UNTIL ZERO.
        JMP LOOP    ;DONE WHEN 0F1 IS ZERO.
        LD A,TON    ;LOAD A WITH Ton RESULT
        X A,00      ;STORE RESULT AT RAM 00
        JMP CONV
.END
```

Figure 4.13 Variable-range PWM software listing.

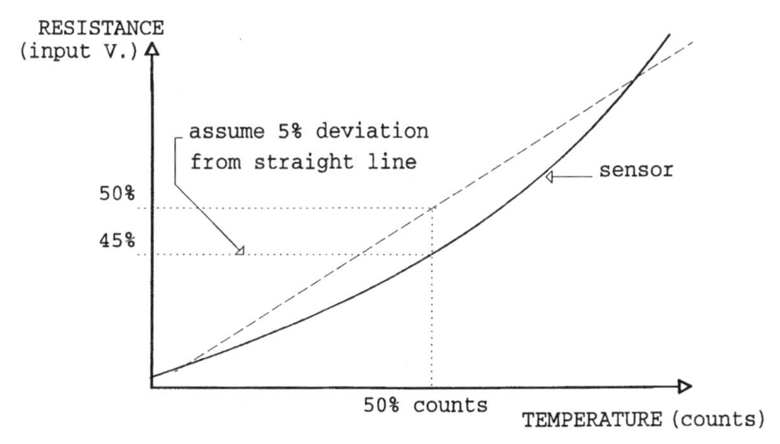

Figure 4.14 Thermistor curve versus temperature.

Nonlinear Measurements

Sometimes it is desirable to measure signals that are inherently nonlinear. Examples include a resistance-temperature device (RTD, or thermistor) and a photodiode detector. In the "typical" thermistor resistance versus temperature curve in Fig. 4.14, note that there is a bow shape to the curve. In this example, there is about a 5 percent deviation at midscale from a straight line connecting the endpoints (for −40 to +85°C).

If this error is undesirable, some form of compensation will be needed to correct for the nonlinearity. Typically, either a large ROM lookup table or a complex mathematical routine is required. Another approach which is simpler and more cost-effective is to alter the durations of the low pulses compared to those of the high pulses. Using a pulse width of 22 cycles for the low pulse and 20 cycles for a high pulse provides a 5 percent reduction in the capacitor voltage at midscale (50 percent of the pulses will be low and 50 percent will be high). This provides a net voltage from the PWM outputs that will follow the input signal profile (see Fig. 4.15) and linearize the A/D counts versus temperature changes. The above approach can be modified to adjust the amount of compensation and the polarity by varying either the high or low pulse durations.

Ordinarily, an amplifier is required to provide gain and offset of the RTD, or thermistor, signal to match the A/D conversion input range. The circuit in Fig. 4.15 will provide a full-range input signal for the A/D converter. However, several additional errors will be introduced with the use of all the resistors, and some calibration will likely have to be made to correct for them. This calibration is usually difficult because of the interaction of the gain and offset adjustments.

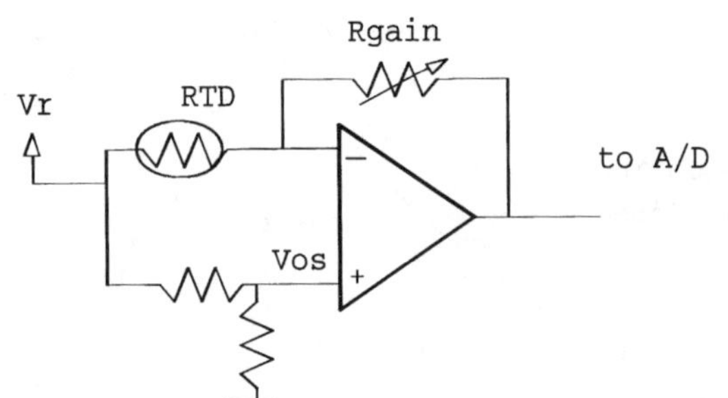

Figure 4.15 Thermistor amplifier.

Note from the PWM profile that the endpoint voltages are determined by the effective reference voltages for the low and high pulses. In other words, the endpoints will converge to the average low-pulse and high-pulse voltages. By using the variable-range technique and modifying the durations as described above, a very simple thermistor circuit can be used (see Fig. 4.16). This simplifies the hardware by eliminating the need for preamplifiers for offset and gain adjustments.

With the below resistor divider circuit, it will be necessary to plot the new profile for change in voltage at the comparator input versus temperature. This is due to the added nonlinearity caused by the combination of a changing resistance of the thermistor R_{th} and the series resistance R_s. More clearly, the total series resistance $R_{th} + R_s$ is not constant in the following equation and will have a nonlinear effect of its own.

$$V_{in} = \frac{R_{th}}{R_{th} + R_s}$$

Potentiometer Measurement Techniques

Many products require an operator input signal to adjust operation. For example, functions such as temperature, position, mode of operation, volume, or light intensity are commonly controlled with a poten-

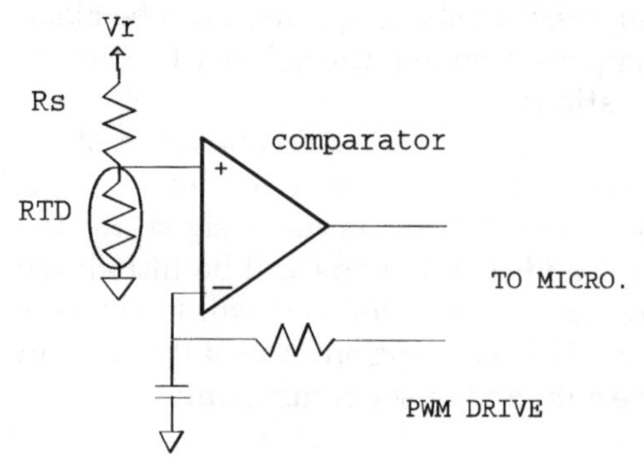

Figure 4.16 Resistor divider with R_s and thermistor.

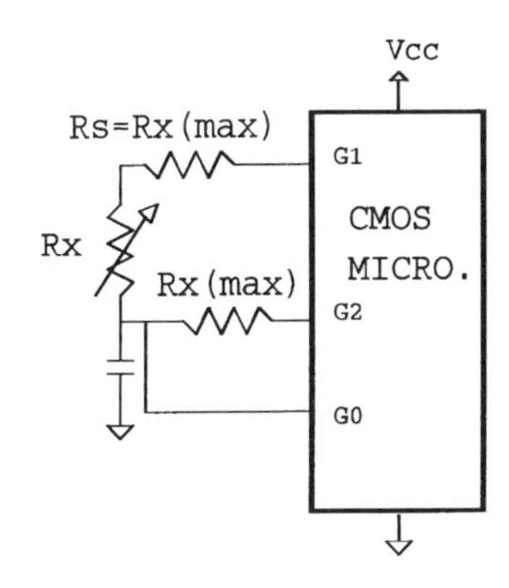

Figure 4.17 Basic potentiometer measurement with microcontroller circuit.

tiometer. By using a very simple circuit with a known resistor, the unknown resistor (or percentage adjustment) can be determined by measuring the ratio of time to either charge or discharge a capacitor. It will be shown that this measurement technique is not sensitive to the tolerances of either the capacitor or the supply voltage. In addition, the timing is linearly proportional to the change in resistance (not exponential, as one might intuitively expect).

Figure 4.17 shows the basic circuit for measuring the unknown resistor R_x. Three connections are required to the microcontroller—one for charging the capacitor with the known resistor, one for charging with the unknown, and one for discharging and then reading the capacitor voltage. The series resistor R_s with R_x is required for limiting the current when R_x is near 0 Ω. Selecting R_s to equal $R_{x(\text{max})}$ makes the mathematics easier to handle and will be explained below.

The mathematical analysis for those skeptical of the linearity is shown below. Determination of the unknown resistance is the same regardless of whether the capacitor is charged or discharged. This is due to the exponential terms falling out of the equation.

Mathematical analysis (basic resistance circuit). Let

V_{il} = microcontroller logic low input threshold

V_{cc} = supply voltage for RC

R_s = known resistance (set equal to maximum R_x)

R_x = unknown resistance

C = fixed capacitance

t_r = delta time with known reference resistance

t_x = delta time with unknown

$R_{\text{eq}} = R_x + R_s$ (total series resistance)

Then

$$V_{\text{cap}} = V_{il} = V_{cc}\left(e^{-\frac{t_x}{R_{eq}C}}\right) = V_{cc}\left(e^{-\frac{t_r}{R_{eq}C}}\right)$$

Solve for t_x by taking the natural log:

$$\ln\left(e^{-\frac{t_x}{R_{eq}C}}\right) = \ln\left(e^{-\frac{t_r}{R_sC}}\right)$$

$$\frac{t_x}{R_{\text{eq}}C} = \frac{t_r}{RC}$$

$$R_{\text{eq}} = R_s\left(\frac{t_x}{t_r}\right)$$

What is usually required is not the actual resistance, but the percentage of R_x (or input). The ratio of t_x/t_r will indicate this percentage. Since R_{eq} can equal up to twice R_x when $R_s = R_{x(\text{max})}$, the ratio of t_x/t_r will be between 1 and 2. This actually helps to simplify the mathematics. Using a 16 × 8 division routine and padding the lower byte of the dividend with 0s eliminate the need for floating-point division. The result will be between 256 and 512. Ignoring the high byte of the quotient (using only the low byte of the result) provides a result of 0 to 255. Zero percent will then equal 00, and 100 percent will equal FF (hexadecimal). For example, let R_x = 50 percent of maximum; then $t_x = 1.5t_r$. Let (measured) t_r = 108 counts; then t_x = 162 counts. *Note:* Padding the low byte is the same as multiplying by 256.

$$256 \times \frac{t_x}{t_r} = 256(1.5) = 384$$

Ignoring the high byte then divides out the 256 factor, and the result will be in the low byte: Result = low byte = 128. *Check:*

$$\%\,(R_x) = \frac{100(384 - 256)}{255} = \frac{(100)(128)}{255} = 50\%$$

Determining R_s

Typically, R_s is set to equal $R_{x(\text{max})}$ for an 8-bit conversion. The resolution can be adjusted, however, by scaling the fixed resistor R_s compared to the maximum unknown resistance. For example, if only 50 counts (2 percent resolution) are required, R_s can be determined from

$$\text{Counts} = \left(\frac{R_{eq}}{R_s - 1}\right)(256) \qquad \text{(basic equation)}$$

$$R_{eq} = R_x + R_s \qquad \text{(total unknown resistance)}$$

For R_x =20 kΩ maximum, solving for R_s gives

$$50 = 256\left(\frac{R_s + R_x}{R_s} - 1\right)$$

$$R_s = \frac{R_x}{0.195} = \frac{20\ \text{k}\Omega}{0.195} = 102\ \text{k}\Omega$$

This example provides a 0 percent reading (00 result in low byte) with R_x = 0 Ω and 100 percent reading (50 decimal) when R_x = 20 kΩ maximum. Since the count is proportional to only the resistance values, the routine will automatically provide a linear result for the values between 0 and 100 percent.

Determining capacitance

With the unknown resistance R_x and the desired resolution known, only the capacitor needs to be determined. Sizing the capacitor can best be described through an example. The example below shows the calculations for an *RC* charging circuit.

Given:

- R_x = 100 kΩ (then R_{eq} − maximum = 200 kΩ)
- Desired resolution = 50 counts
- Single count duration for polling = 10 µs
- V_{ref} = 5 V
- $V_{ih(\text{micro})*} = (V_{cc})(0.7) = (5\ \text{V})(0.7) = 3.5\ \text{V}$ (a 0.7 factor is typical)

Determine *C*.

$$\text{Total time for measurement} = 10\ \mu\text{s} \times 50 = 500\ \mu\text{s}$$

$$V_c(t) = V_{ref}\left(1 - e - \frac{t}{RC}\right)$$

*(micro) = microcontroller.

$$3.5 = 5\left(1 - e - \frac{500\ \mu s}{200\ kC}\right)$$

$$\ln\left(1 - \frac{3.5}{5}\right) = -\frac{500\ \mu s}{200\ kC}$$

$$C = \frac{500\ \mu s}{200\ k\Omega \times 1.2} = 2.08\ nF \qquad \text{Use } 0.002\ \mu F.$$

Keep in mind that the component tolerances can reduce the effective resolution by shortening the timing duration. The largest contributing tolerance to consider is the input logic thresholds of the microcontroller. To compensate for this, it is recommended that you set the resolution (counts) to the required amount for the worst-case (minimum) timing cycle. In other words, it is necessary to pad the resolution to compensate for the circuit tolerances which will shorten the timing cycle.

Software description

Either a software polling or a hardware timer with input capture can be used for timing the charge/discharge cycle. Polling (software timing the capacitor ramp voltage level) is simple, but it can have some drawbacks. Among the drawbacks are the increased time for counting and the requirement that all interrupts be disabled during counting. Conversely, there are two benefits from using the input capture timing: The speed of the conversion is increased, and other functions can be performed while the timing cycle is under way, thus eliminating the possibility of unwanted interrupt latency.

Using a microcontroller with software-programmable I/O and the capability for tristating will simplify the hardware and software. Otherwise additional I/O and some transistors will be necessary for tristating. The basic flowchart for the polling method is shown in Fig. 4.18. Care should be taken when the potentiometer is at or near either 0 or 100 percent. Due to the tolerance of the circuit resistances, it is possible (and likely) that an underflow or overflow can occur within the mathematics routine. This can be easily dealt with by checking the high byte of the result. For instance, when $R_s = R_{x(max)}$, the result should be between 256 and 512. This means that the high-byte bit 0 should be set for all measurements from 256 (0 percent) up to 511 (near 100 percent). If neither bit 0 nor bit 1 is set, then an underflow has occurred and the result is 0 percent. The same idea can be applied for checking an overflow. If any bit other than bit 0 is set in the high byte, 100 percent is indicated.

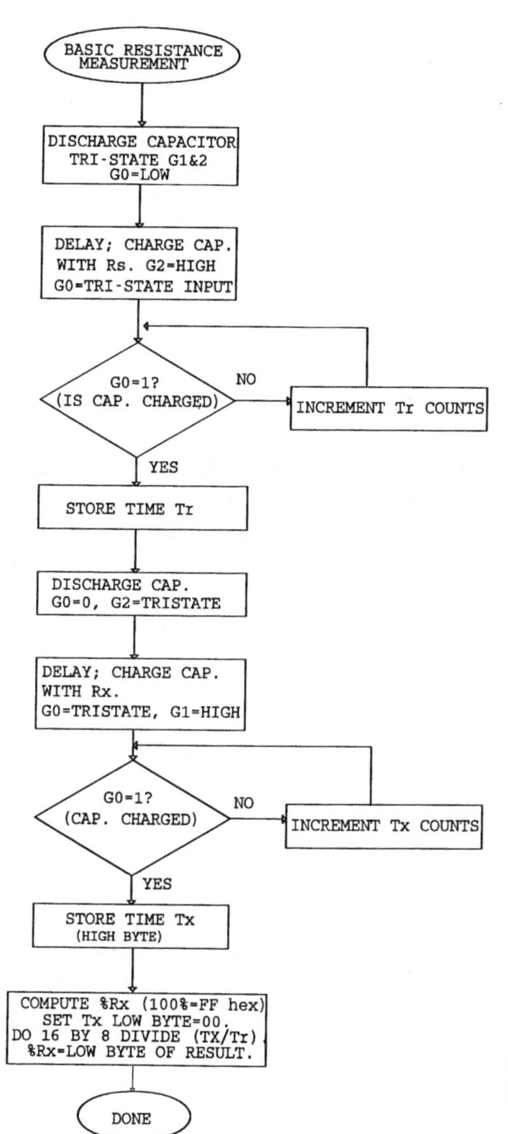

Figure 4.18 Basic flowchart of the polling method.

Error Analysis

The logic thresholds (either high or low) and the supply voltage are not error factors. This is due to the fact that the same levels are used for both the reference and the unknown measurements. The actual value or tolerance of the capacitor is not critical either, since the same capacitor is used for both measurements and only the ratio of times is used. However, care must be taken to ensure that the capacitor is either precharged or discharged to the same voltage for each measurement. Otherwise the difference in counts will show up as error. Given proper design, the accuracy is limited only to the tolerance of the unknown and fixed resistances. Since potentiometers are typically used where an operator can visually adjust to the desired setting, only the resolution is likely to be important.

Multiple Potentiometer Measurements

When multiple potentiometers must be measured and it is undesirable to tie up three microcontroller pins for each, the following technique can be used. In the circuit in Fig. 4.19, the diodes in each leg of the circuit allow isolation when the anode side is set to ground.

The value of resistor R needs to be high enough that the current is limited at the microcontroller I/O pins. This is most important when the potentiometers are near zero Ω and I/O pins are reset low. Furthermore, the series resistor R will limit the current through the microcontroller during clamping when $+V>V_{cc}$. Note in Fig. 4.19 that more potentiometers can be added in parallel, thus requiring only one additional pin for each.

Figure 4.20 is a flowchart for performing multiple resistance measurements. This is very similar to the basic idea, but requires the potentiometers to be individually selected by driving the diodes with the proper logic level.

Note that the potentiometers are referenced to V_{supply} instead of both ends to the microcontroller. This is not necessary, but it illustrates another way of configuring the circuit. Another adaptation of the circuit is shown in Fig. 4.21, where the potentiometers are referenced to ground. This is often the case in automotive applications. The procedure for this is very similar to that used earlier. Instead,

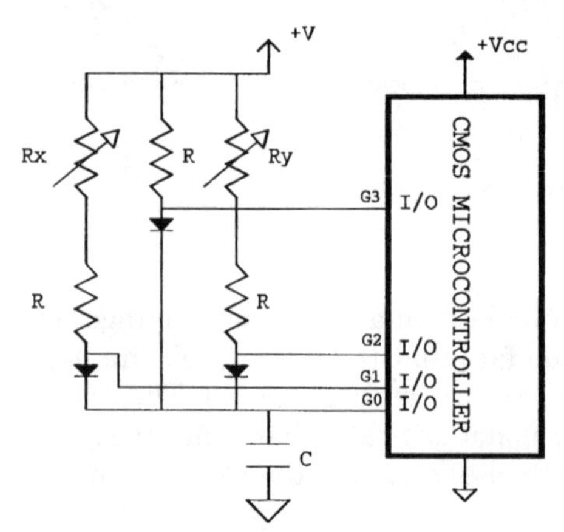

Figure 4.19 Multiple-potentiometer circuit referenced to V_{supply}.

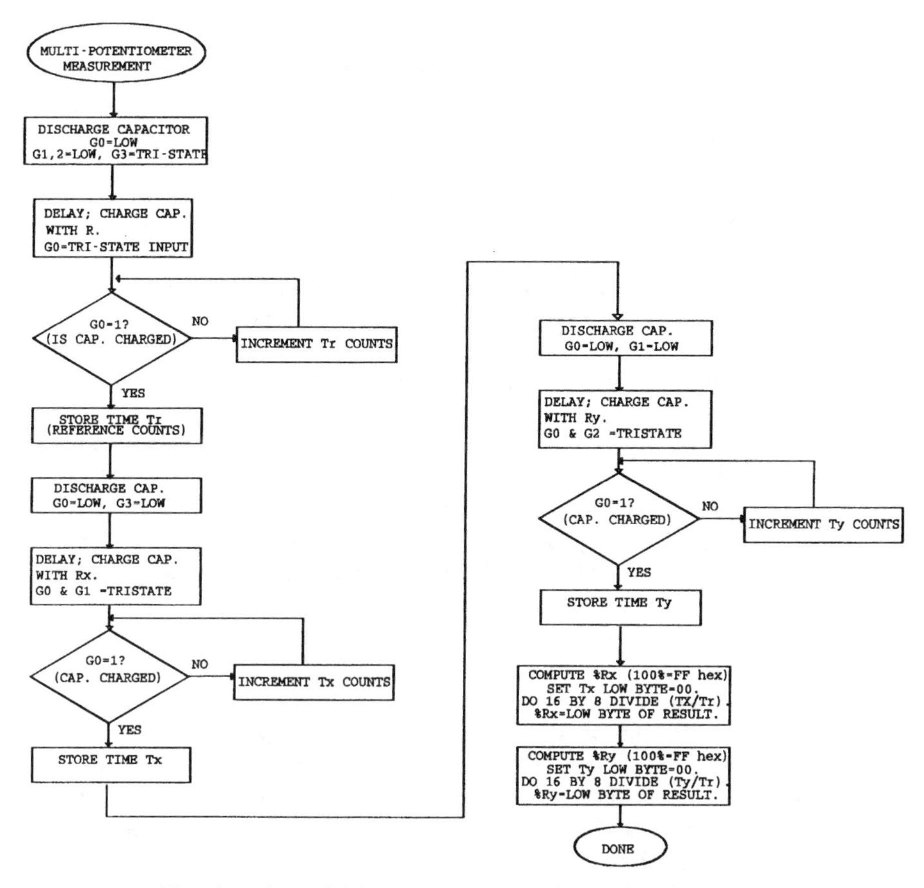

Figure 4.20 Flowchart for multiple potentiometers referenced to V_{ref}.

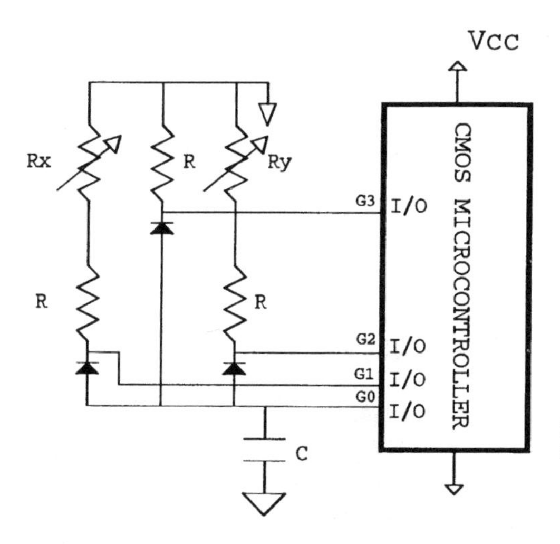

Figure 4.21 Ground-referenced multiple-potentiometer circuit.

the capacitor voltage is timed during a discharge cycle, and the cathode side of the diodes must be driven to V_{cc} for isolation.

Dealing with Low-Value Resistances

In some situations the values of the circuit resistances may become a loading problem for the microcontroller. It is very important that the initial capacitor charge voltages be the same for all measurements. Due to limited drive capability with microcontrollers, low-value resistors may cause unwanted errors by not providing sufficient blocking voltage for the diodes. This problem can be overcome by using an inexpensive transistor for buffering, as shown in Fig. 4.22.

Another problem related to using low-value resistors can result in the need to increase the value of the capacitor. Since sufficient time must be given for the microcontroller to count up to the maximum resolution, the *RC* time constant must be long enough. Microcontrollers typically have limited drive capability and require significant resistance values to limit current. Two options exist for solving this problem. Either the drive current for charging the capacitor can be increased with the aid of an extra transistor, or more time will be required to fully charge the capacitor. The drawback to using the transistor approach, besides the extra cost, is that another microcontroller pin will be required for monitoring the capacitor logic level. If possible, the preferred method is to precharge the capacitor in the software routine well before the resistances are measured. This reduces the total conversion time and possible interrupt latency. When several resistances must be measured, measuring only one per software loop facilitates the use of the capacitor precharge method.

R-2*R* A/D Converter

Successive approximation can be accomplished very cost-effectively by using an *R*-2*R* thick-film network. These networks, described in Chap. 2, can provide up to 10 bits of resolution. The circuit in Fig.

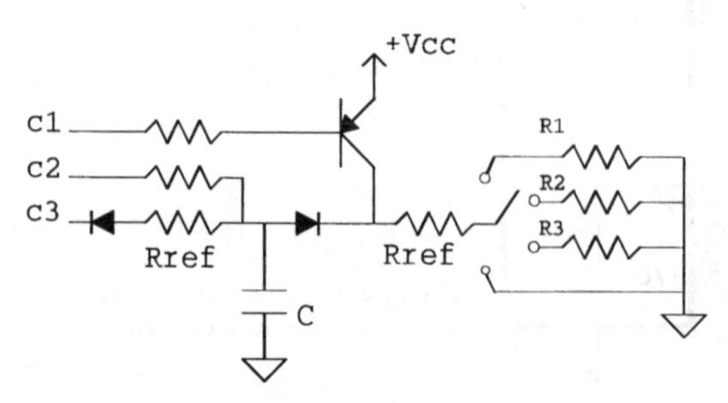

Figure 4.22 Transistor buffer circuit.

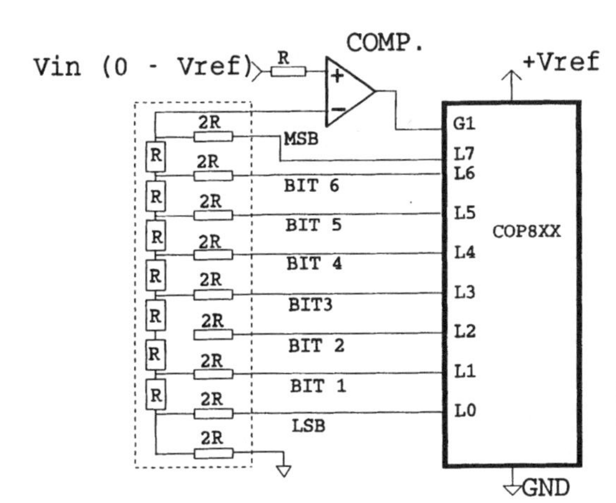

Figure 4.23 An R-$2R$ A/D basic circuit.

4.23 shows the required components and I/O pins for an 8-bit conversion. Note that there is a drawback with this approach in that $n+1$ microcontroller pins are required, where n is the number of bits of resolution. This problem may be outweighed by the advantage of a relatively faster conversion, in the area of 80 instruction cycles.

Just like the dedicated hardware version of the successive-approximation converter, each bit is set one at a time. With a CMOS microcontroller that switches very close to the supply and ground, the error associated with driving an R-$2R$ network of 100 kΩ or more will be negligible. First, the most significant bit (MSB) is set, and the comparator will determine if the bit should stay set. If the voltage produced by the R-$2R$ network drives the comparator low, this indicates that the estimate was too high and should be reset. Each successive bit test will provide a more accurate estimate of the input until the LSB is tested. The software flowchart and code for the COP800 microcontrollers are shown in Figs. 4.24 and 4.25, respectively.

Voltage-to-Frequency Converters

One method to achieve high resolution in measuring either a variable resistance or capacitance sensor is to use a voltage-to-frequency (V/F) conversion. The V/F converter will alter the output frequency proportionally to changes in the sensors. Another method is to use an input voltage to vary the frequency (voltage-controlled oscillator). These methods provide a relatively high-resolution and high-speed conversion, but will suffer from the same tolerance problems as the RC time constant method, with the initial and temperature tolerance of the capacitor being the most severe error.

However, there are some advantages to using a voltage-to-frequency converter. It is not difficult to measure the output frequency with

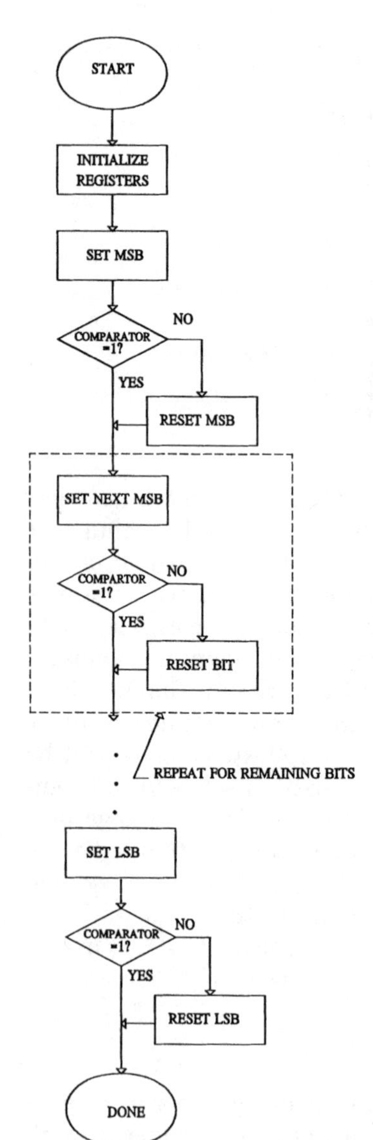

Figure 4.24 The *R*-2*R* flowchart.

high resolution by using a timer on board a microcontroller. This can be accomplished with a single input connected to the timer set up for input capture mode. In this mode, the timer can automatically store the time between pulses from one edge to the other.

The major drawback to the V/F converter is the difficulty of obtaining accuracy without some calibration. However, some product designs will require calibration regardless of the A/D conversion method. An example is the common household digital scale. These measurements are based on either a resistance or a capacitance

```
;THIS PROGRAM PERFORMS AN 8 BIT A/D CONVERSION WITH THE USE OF AN
;R-2R NETWORK (SUCCESSIVE APPROXIMATION) AND CONVERTS IN LESS THAN 80 CLKS.
;Vin IS ON THE + COMP. INPUT, DAC OUTPUT IS ON COMP - INPUT, COMP. OUT GOES TO
;G1 ON COP8XX.  THE R-2R NETWORK IS DRIVEN BY THE COP L PORT WITH LO DRIVING
;THE LSB AND CONTINUING TO L7 DRIVING THE MSB.
.CHIP 820
;       INITIALIZATION
LD 0FE,#0D0 ;LOAD B REG TO POINT TO L PORT
LD 0D0,#080  ;LOAD L PORT DATA REG WITH MSB
LD 0D1,#0FF ;SET L PORT AS OUTPUTS
SBIT 1,0D4 ;G1 AS PULL-UP
RBIT 1,0D5 ;G1 AS INPUT
;
;       TEST MSB
IFBIT 1,0D6 ;TEST COMPARATOR OUTPUT
JP .+2        ;JUMP IF COMP.=1
RBIT 7,[B]  ;OTHERWISE RESET MSB IF COMP.=0
;
;       TEST BIT 6
SBIT 6,[B]
IFBIT 1,0D6
JP .+2
RBIT 6,[B]
;
;       TEST BIT5
SBIT 5,[B]
IFBIT 1,0D6
JP .+2
RBIT 5,[B]
;
;       TEST BIT4
SBIT 4,[B]
IFBIT 1,0D6
JP .+2
RBIT 4,[B]
;
;       TEST BIT3
SBIT 3,[B]
IFBIT 1,0D6
JP .+2
RBIT 3,[B]
;
;       TEST BIT2
SBIT 2,[B]
IFBIT 1,0D6
```

Figure 4.25 *R*-2*R* code listing.

change when the weight is placed on the scale. Calibration is required due to the initial tolerance of the sensors and other circuit components. Temperature drifts would be the only major concern following calibration, and they can usually be controlled to acceptable levels with the proper choice of film capacitor (i.e., polycarbonate).

Practical V/F circuits

One practical circuit for performing a low-cost V/F conversion utilizes the ever-popular LM555 timer. This circuit can use a variable resis-

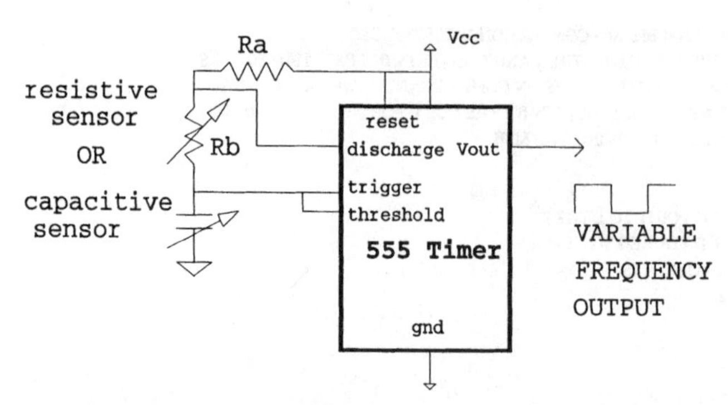

Figure 4.26 The LM555 V/F circuit.

tance or capacitance, as shown in Fig. 4.26. One example of using a capacitor sensor is for a weighing scale. Variable resistances can indicate functions such as temperature (i.e., thermistors) or set points (i.e., potentiometers).

Operating in the astable mode (free-running oscillator), the LM555 will continuously retrigger itself by charging up to two-thirds the supply voltage V_{cc} and then discharging to $\frac{1}{3}V_{cc}$. Therefore it is independent of changes in the supply voltages. The following equations describe how the output frequency can be used to indicate the unknown parameters.

The charge time interval is determined by

$$t_c = 0.693(R_a + R_b)C$$

The discharge time interval is determined by

$$t_d = 0.693R_bC$$

Therefore the output frequency is

$$\text{Frequency} = \frac{1}{t_c + t_d} = \frac{1.44}{(R_a + 2R_b)C}$$

Note that the output frequency is linearly proportional to changes in the capacitor C, but nonlinearly proportional to changes in the resistance. This is due to the frequency changing, with one of the two resistors being constant. Full-scale measurements require that the maximum resolution be achieved with the microcontroller timer. This will be limited by the timer clock and the resolution, as described below.

$$\text{Maximum frequency measured} = \frac{1}{\text{timer clock} \times \text{resolution}}$$

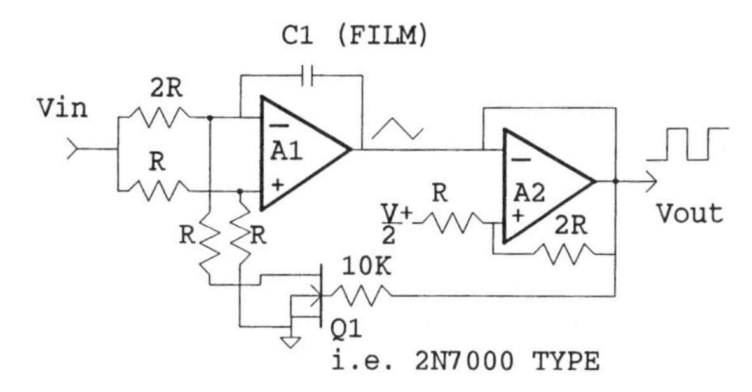

Figure 4.27 Two-stage amplifier V/F circuit.

An alternate method of creating a V/F converter is to use a two-stage amplifier circuit, as shown in Fig. 4.27. This circuit will vary the output frequency with changes in either the input voltage or *RC*. The first stage provides a triangular waveform by integrating half the input voltage. This triangular waveform will have a period that is proportional to the input voltage. The second stage is configured as an oscillator. Each time the square wave from the second stage goes high, the transistor will change the integration polarity to a positive-going ramp by switching in a divider circuit. When the triangular waveform reaches $\frac{2}{3}V_{supply}$, the output will go low. This releases the transistor divider and starts the ramp going in the negative direction. When the triangle reaches $\frac{1}{3}V_{supply}$, the process is repeated.

Depending on the desired accuracy (i.e., drift) and range, the amplifier can be chosen to meet specific needs. For best operation, a dual CMOS amplifier such as the LMC6482 can be used. This amplifier will reduce the errors by switching the output very close to ground on the second stage, therefore providing more accurate thresholds for the oscillator. In addition, the input range of the LMC6482 includes the power supply rails for increased input range, and the input leakage current will not affect accuracy. Further improvement can be made by using a FET switch (i.e., 2N7000) for the reset action and a resistor network for tighter tolerance tracking.

When you are selecting the amplifiers, keep in mind that the amplifiers will have limited bandwidth, as described below:

$$\text{Bandwidth} = \frac{\text{slew rate}}{6.28 \times V_{p-p}}$$

Also keep in mind that the low end of the input voltage range will be limited by the total input offset voltage caused by the integrating amplifier circuit.

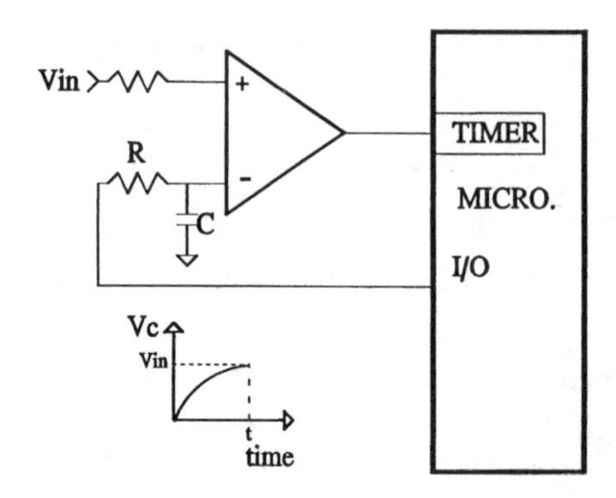

Figure 4.28 *RC* circuit, comparator, and uc.

RC Charge Time Constant A/D Converter

This very simple technique can be considered a crude single-slope A/D converter. It is based on the time to charge a capacitor from one level to another voltage reference level at a comparator, as shown in Fig. 4.28. By using a known capacitor, reference voltage, and a known resistor, the time for the exponential ramp to reach the unknown input voltage indicates the unknown input voltage level. Although this is very simple in principle, there are several complications and potential drawbacks, the most obvious of which is the nonlinear timing slope. Tight-tolerance components are also required to predict the curve with any reasonable accuracy. Tight-tolerance resistors are not expensive, but capacitors with tight tolerance over temperature are. Either a polycarbonate or a polystyrene capacitor is a good choice for a stable measurement over temperature. The reference voltage tolerance (if not ratiometric) and comparator input offset voltage are other possible sources of error.

The charging nonlinear timing function is described by

$$V_c(t) = V_0 + (V_{\text{ref}} - V_0)\left(1 - e^{-t/RC}\right)$$

One practical method of achieving a conversion with this circuit is to use a ROM lookup table. Nominal component values for *R, C,* and V_{ref} should be used to calculate values of counts (input voltage) versus time from the above equation. One other difficulty with timing is due to the exponential slope. Note that there will be less resolution in the beginning of the slope, where it is steep, compared to the tail end. Additionally, there will be significantly more time to complete a conversion for $V_{in} >$ approximately $0.9V_{\text{ref}}$ where the slope is relatively flat.

Another method of dealing with the nonlinear slope is to use a piecewise linear approximation, as shown in Fig. 4.29. With several straight

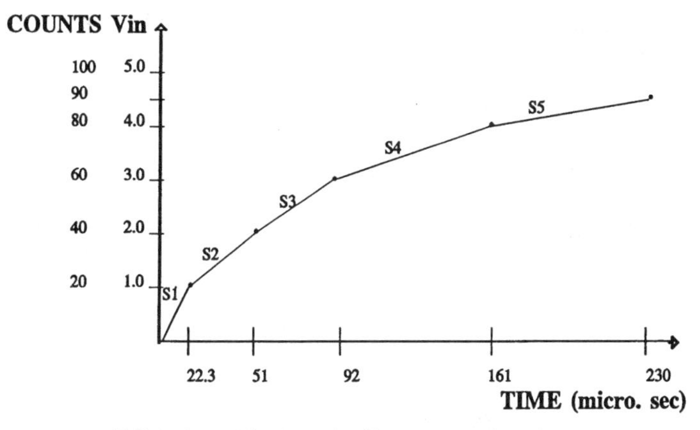

Figure 4.29 *RC* timing with piecewise linear approximation.

lines approximating the exponential slope, all that is needed is to store the time breakpoints and the corresponding slopes. The slopes can be described by easy ratios of counts over time. With a constant denominator of 256, the mathematics for computing the compensated resultant counts is made very simple. All that is needed is to perform an 8 × 8 multiplication and throw away the low byte (the same as dividing by 256). Using a large value for the denominator also increases the accuracy of describing the slope with integers. The following example illustrates how to design an *RC* charge time constant circuit.

***RC* A/D converter example.** Let

$$RC = 100\ \mu s$$

$$\text{Maximum counts} = 100 \qquad (1\%\ \text{resolution})$$

$$V_{ref} = 5\ V$$

$$\text{Minimum clock cycle} = 1\ \mu s$$

First, solve for t with counts (volts) = 1, 2, 3, 4, 4.5 V. Note that the output counts and corresponding voltages are in a linear scale in Fig. 4.29. From

$$V_c(t) = V_{ref}\left(1 - e^{-t/RC}\right)$$

$$t = RC \ln\left(1 - \frac{V_c}{5}\right)$$

Substituting V_c = 1, 2, 3, 4, and 4.5 gives

$t_1 = 22.3\ \mu s \qquad t_2 = 51\ \mu s \qquad t_3 = 92\ \mu s \qquad t_4 = 161\ \mu s \qquad t_5 = 230\ \mu s$

Second, compute the delta times $[T(i) = T_i - T(i - 1)]$ between the above breakpoints.

$$T_1 = 22.3\ \mu s \qquad T_2 = 51 - 22.3 = 28.7\ \mu s \qquad T_3 = 92 - 51 = 41\ \mu s$$

$$T_4 = 161 - 92 = 69\ \mu s \qquad T_5 = 230 - 161 = 69\ \mu s$$

Third, compute the individual slopes S from the ratio of counts to delta time.

Slope	Approximation
$S_1 = \frac{20}{22.3}$	$= \frac{230}{256}$
$S_2 = \frac{20}{28.7}$	$= \frac{178}{256}$
$S_3 = \frac{20}{41}$	$= \frac{124}{256}$
$S_4 = \frac{20}{69}$	$= \frac{74}{256}$
$S_5 = \frac{10}{69}$	$= \frac{37}{256}$

From the above information and the following equation, the output counts can be computed. Refer to Fig. 4.30 for an example of a software flowchart.

$$\text{Output counts} = C_i + (t - t_i)S_i$$

where C_i = output counts at corresponding breakpoints
t = actual timing counts
t_i = time at next lower breakpoint from t
S_i = approximate slope at corresponding breakpoint, expressed as output counts per time

Note: Variables C_i, t_i, and S_i are constants for segments 1 to 5 stored in ROM.

Suppose that $t = 35\ \mu s$; determine the calculated output counts. The next-lower breakpoint from 35 µs in the lookup table is $t_2 = 22\ \mu s$ with $C_2 = 20$ and $S_2 = 178$.

$$\text{Counts} = 20 + \frac{(35 - 22)178}{256} = 29$$

Note: Performing 8×8 multiplication and ignoring the low byte are

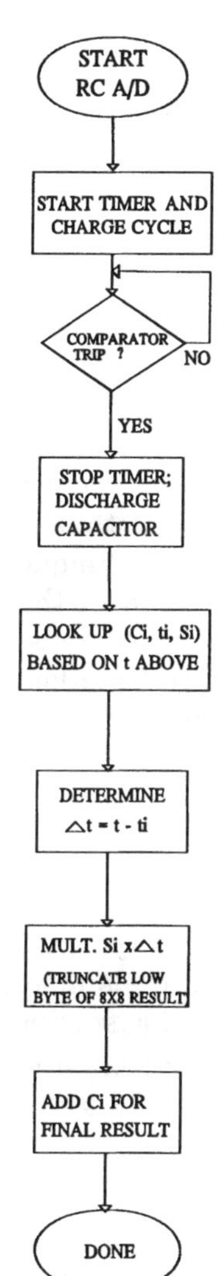

Figure 4.30 Flowchart for *RC* time constant A/D converter.

the same as automatically dividing by 256. The example error (percentage of full scale) is

$$V_c(t = 35) = 5(1 - e^{-35/100}) = 1.476 \text{ V}$$

$$V_c \text{ calculated} = (5 \text{ V})(29/100) = 1.45 \text{ V}$$

$$\% \text{ full-scale error} = \frac{100(1.476 - 1.45)}{5}$$

$$= 0.5$$

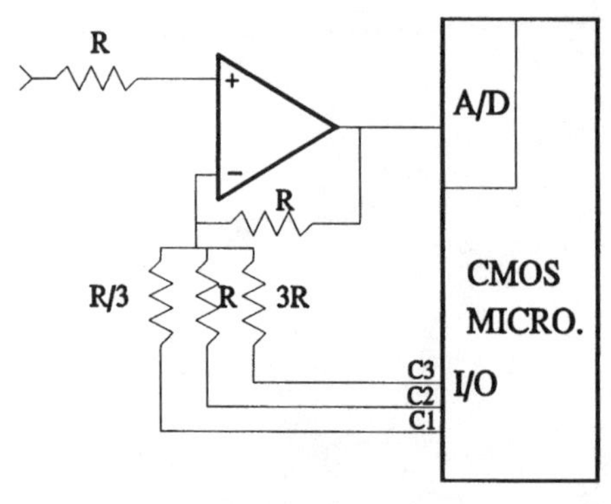

Figure 4.31 Variable-gain microcircuit.

Providing that resolution, not high accuracy (typically more than 8 percent error with reasonable tolerance components), is all that is really required, this approach can serve your needs. This technique provides relatively fast conversions (limited by the speed of the counter) and uses little program ROM or external hardware. However, the only real advantage of the *RC* time constant method compared to the PWM A/D conversion is speed, with the improved PWM A/D conversion superior in every other category.

Microcontroller On-Board A/D Converter Enhancements

Some applications require higher resolution than the 8 bits typically available on board the microcontroller. As long as resolution is more important than accuracy, there are two low-cost options for solving this problem. These involve varying either the gain of the signal or the reference voltage (using the differential mode) based on a previous measurement.

Variable gain

By using an amplifier that can vary the input signal gain as shown in the example circuit in Fig. 4.31, the effective resolution can be increased from 8 to 10 bits. One procedure is to switch in the highest gain on the amplifier and do a trial conversion. If the result is less than 255, then the input was in the first quadrant (2 MSBs = 00). Otherwise, the next-lower gain is tried until the result is less than the maximum of 255. With each successive try, the MSBs will have to be incremented by 1.

Variable reference

An alternative to the above approach is to use the differential mode of the on-board A/D converter and use multiple references for the V(−)

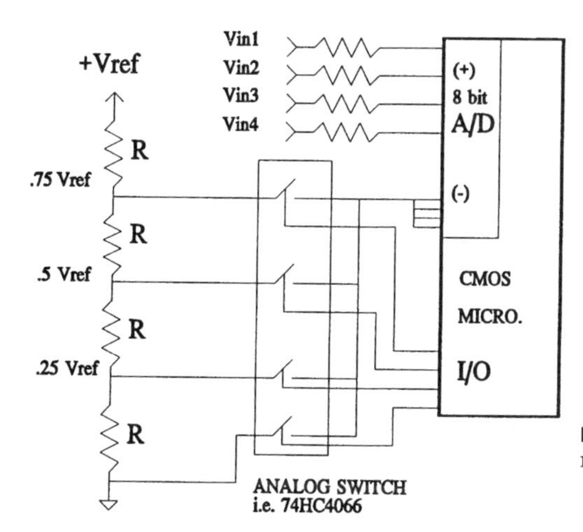

Figure 4.32 Variable-reference microcircuit.

inputs. For example, the circuit in Fig. 4.32 shows four equal resistors that provide increments of one-fourth the full range. This is very similar to the previous procedure and works by dividing up the input range with multiple references instead of multiple external gains.

The procedure is begun by doing a conversion on the channel with the highest reference (i.e., in this example $3V_{ref}/4$) and then checking a result less than 255. Since $\frac{3}{4}V_{ref}$ was subtracted from each input, the first measurement will check the range of $V_{ref} - \frac{3}{4}V_{ref}$. If the result is less than 255, then the input is in the first one-quarter range, and the 2 MSBs are reset to 00. Otherwise, the next-lower reference $\frac{1}{2}V_{ref}$ is tried, and if a result less than 255 occurs, the 2 MSBs are set to 01. The process continues until the results are less than 255. When a result between 0 and 255 is produced, the 8-bit conversion represents the lower byte of the total result, and the selected range where the result occurred indicates the 2 MSBs.

As can be seen in the example circuit in Fig. 4.32, one drawback is the reduction of available A/D input channels. This is due to the use of multiple channels in the differential mode. Among the advantages of this approach are the elimination of the external programmable amplifier(s) and an overall increase in accuracy with the analog switch.

Chapter

5

Single- and Multislope A/D Conversion Techniques

Single- and multislope A/D converters are appropriate for very high-accuracy or high-resolution measurements where the input signal bandwidth is relatively low. Besides accuracy, single- and multislope converters offer a low-cost alternative to other A/D conversion techniques such as the successive-approximation approach. Typical applications for single- and multislope converters include digital voltmeters, weighing scales, and process control. These types of converters also often find applications in battery-powered instrumentation due to the capability for very low power consumption.

As the name implies, single-slope A/D converters use only one ramp cycle to measure each input signal. Multislope converters, on the other hand, require at least two (dual-slope) or more (multislope) ramp measurements for each input. Both techniques are somewhat similar in that an integrating function is vital for measuring the input signals. There are some subtle differences between the single- and multiple-slope A/D converters. Generally, single-slope converters can be used for up to 14-bit accuracy, while multislope converters are preferred for accuracy between 14 and 25 bits. This is because multislope converters are less susceptible to noise than single-slope converters.

Noise is reduced in multislope converters because the input signal (including noise) is integrated. This differs from the single-slope converter that generates a fixed ramp for comparing with the input signal. Therefore, any system noise present at the comparator input while the ramp is near the threshold crossing can cause errors. For obvious reasons (one versus multiple slopes), a single-slope converter is faster than a multislope converter for the same resolution. Another performance difference is that hardware errors are inherently canceled in the single-slope converter. Conversely, the multislope converter must use an extra cancellation technique. Multislope converters also have to contend with dielectric absorption in the integrator capacitor. Due to these difficulties in eliminating errors with multislope converters, the single-slope conversion method is considerably less expensive. Depending on the application, each type has its own advantages.

The design examples in this chapter are intended to point out many useful ideas, and they should stimulate several other alternatives. First, single-slope converter techniques are discussed and supported with design examples. Then several multislope A/D conversion techniques are explained and supported with a strain gage application. This example also illustrates many other aspects of analog circuit design.

Single-Slope A/D Converter

The basic idea behind the single-slope converter is to time how long it takes for a ramp to equal an input signal at a comparator. Absolute measurements require that an accurate reference (V_{ref}) matching the desired accuracy be used for comparing the time with the unknown input measurement. In other words, the unknown input (V_{un}) can be determined by

$$V_{un} = V_{ref}\left(\frac{t_{un}}{t_{ref}}\right)$$

where t_{un} = counts for measuring V_{un} and t_{ref} = counts for measuring V_{ref}. Ratiometric conversions do not require a reference at all. These measurements require only the ratio of two or more measurement counts (that is, t_1/t_2), where the ratio is directly proportional to the difference in magnitudes.

The heart of the single-slope analog-to-digital converter is the ramp voltage required to compare with the input signal. Providing that this ramp function is highly linear, the system errors will be completely canceled. This is because of the way the input signals are measured. Since each input is measured with the same ramp signal and hardware, the component tolerances are exactly the same for each measurement. Regardless of the initial conditions or temperature drifts, no calibration or autozero function is required.

Ramp generation

There are several ways to build a ramp generator for use in the single-slope A/D converter. The ramp function can be made to go in either a positive or a negative direction for measuring purposes. Positive-going ramps are easier to deal with than negative-going ramps. This is due to the extra operation necessary in software for subtracting the difference between V_{ref} counts and V_{in} counts on every measurement. When a ramp circuit is designed, the following characteristics should be considered:

Ramp circuit criteria:

1. Linearity
2. Output $V_{p\text{-}p}$ voltage
3. *dv/dt* (ramp time from 0 to 100 percent)
4. Cost

Among the available options are a discrete circuit with transistors, an adjustable regulator, and an amplifier-based design. If you are considering using an integrated current source, be careful to check linearity and price. Figure 5.1 illustrates two simple low-cost ramp generator circuits. The circuit in Fig. 5.1*a* works by providing a constant current source based on Vd/R for driving a capacitor. With a small signal transistor, or CMOS logic gate, the capacitor can be discharged prior to starting a measurement. When the transistor is turned off, the current will start generating the linear ramp with the capacitor.

In Fig. 5.1*b*, the low-cost LM317L adjustable regulator is configured as a current limiter or, more accurately, a constant current source. This circuit works by maintaining a constant + 1.2 V across R_s by

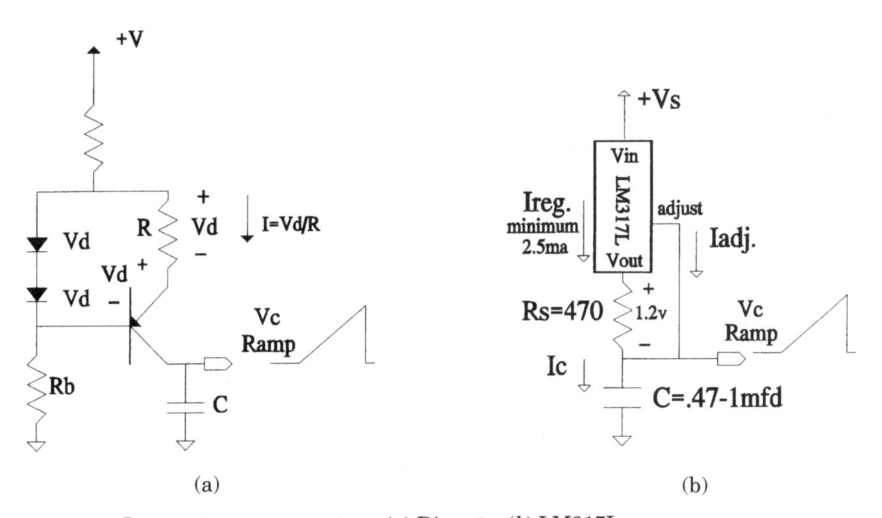

Figure 5.1 Low-cost ramp generators: (*a*) Discrete; (*b*) LM317L.

adjusting the differential voltage between V_{in} and V_{out}. When the discharge transistor is actively sinking 2.5 mA, the ramp (and capacitor) voltage is at zero. As soon as the transistor is turned off, the 2.5-mA current will start filling up the capacitor for generating a linear ramp. There are a few things to be aware of when you use the adjustable regulator in this mode. First, it is important to maintain a minimum load of 2.5 mA (with $V_{in} - V_{out} < 15$ V) for good regulation (linearity). Second, it is necessary to maintain a minimum of 2 V across V_{in}/V_{out} with a light load current over the full operating range. Since the LM317 I_{adj} stays relatively constant during a single measurement and it is much smaller than the charge current, there will be no resulting error. The line regulation, however, will cause some linearity error. As the voltage across the LM317L changes, the current can change by a maximum of 0.07 percent per volt (worst case) over the full operating temperature of − 40 to + 125°C. This means that the linearity of the circuit can exceed 9 bits (3 × 0.07 percent = 0.21 percent).

For higher accuracy (i.e. linearity), the circuit in Fig. 5.2 can provide a highly linear ramp. This utilizes an amplifier configured as an integrator with the capacitor in the feedback. The trick to making this work with a single supply is to bias up the positive amplifier input so that the ramp can be easily controlled with logic levels. When the control is low, the integrator will produce a positive-going output ramp. With a high control signal, the ramp will be driven in a negative direction.

Integrator amplifier selection

When the amplifier is chosen, it is important to note the input bias current I_b. If the input bias current or leakage current is nearly the same magnitude as the charge current I_{ch}, it can affect the ramp linearity, although this is not critical as long as the leakage current stays constant during one complete measurement cycle.

For optimum performance, a CMOS amplifier such as the LMC6034 from National Semiconductor can be used. This type of

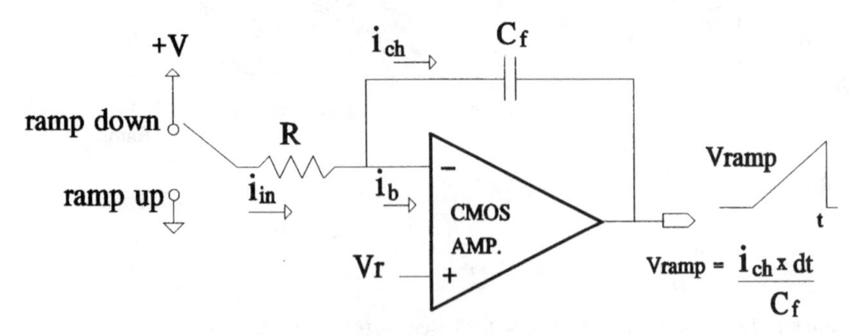

Figure 5.2 Integrator ramp generator.

amplifier has a negligible worst-case input leakage current of 200 pA at 85°C. Another advantage of the LMC6xx family is that the outputs can drive to within about 10 mV of the supplies. This makes it easier to work with since the output will be linear for nearly the entire supply range. Depending on the desired input range, it may be possible to avoid having to bias up the input signals prior to going into the comparator. In less critical applications, a standard bipolar amplifier such as the LM324 with an input bias current as high as 250 nA can be an option. It is still possible to achieve very good performance, providing the charge current is set much higher than the input bias current I_b.

Sizing the capacitor

As mentioned above, the charge current should be set much greater than the circuit leakage current. This includes not only the amplifier (or transistor) but also the circuit board and capacitor leakage current. The leakage current on the capacitor has to be very low and can be easily eliminated by selecting a film capacitor such as a polypropylene one. Circuit board leakage should not be a major concern except for high humidity, in which case a guard ring around the input pins tied to the reference voltage or an epoxy coating will significantly reduce the effects. Chapter 9 discusses methods of creating guard rings and other circuit layout techniques.

In order to set the capacitor value, it will be necessary to determine the following:

1. Time to count maximum resolution dt
2. Required ramp output $dV_{\text{p-p}}$
3. Desired charge current I_{ch}

Given the above, the capacitor can be computed as follows:

$$\text{Capacitance} = I_{\text{ch}} \frac{dt}{dV_{\text{p-p}}}$$

Single-slope design example

As the first example, we look at designing a 12-bit single-slope A/D converter. Assume that the input range is 0 to 5 V and that it is desired to measure six input signals, some of which need to be measured against a 5-V reference for an absolute measurement. The circuit in Fig. 5.3 meets the desired requirements. This circuit utilizes a microcontroller for switching up to six input channels, controlling the ramp generator, and timing the charge cycles. Even if the system did not already require a microcontroller, dedicating one for this purpose

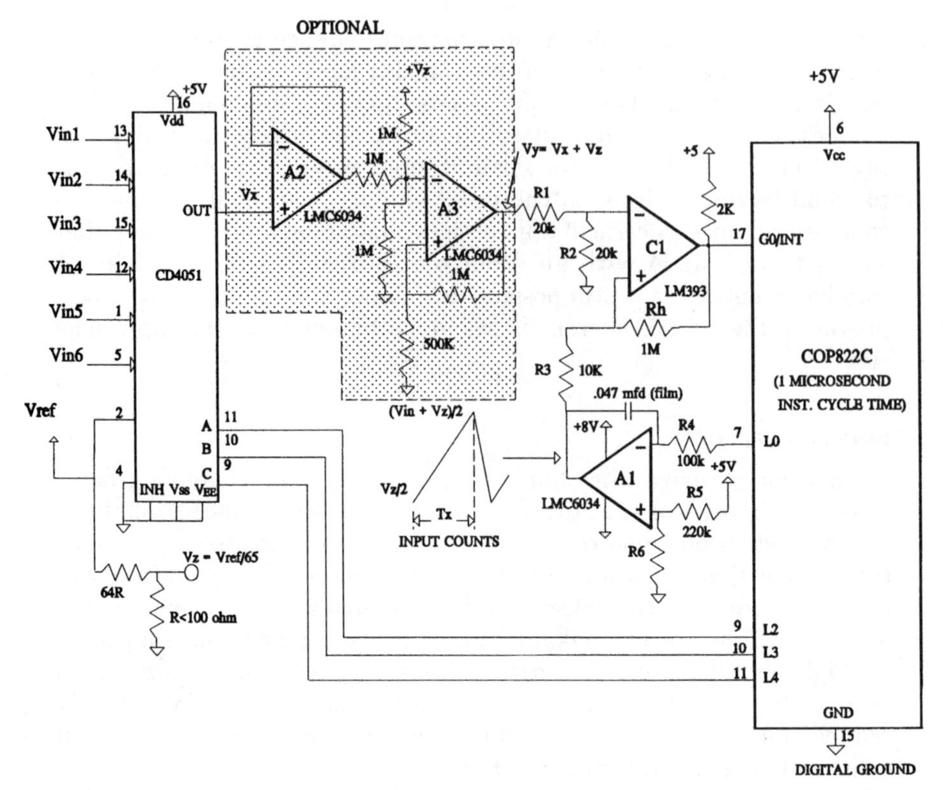

Figure 5.3 Example of a 12-bit single-slope design.

might still be desirable. The microcontroller will require very little memory, I/O, or special features to perform the A/D conversion. Therefore, a very cost-effective microcontroller can be selected, and there will still be plenty of reserve power for other tasks, if necessary.

When the input capture method is used for timing the comparator edges, some hysteresis is necessary. This is provided with R_h and R_3 as shown in Fig. 5.3.

$$V_h = V_{cc}\left(\frac{R_3}{R_3 + R_H}\right)$$

Any system noise present near the comparator threshold will potentially cause multiple edges to be captured in the timer.

The ramp generator is created with one of four amplifiers from the LMC6034. This is an extremely high-input-impedance, high-gain amplifier with an output swing that goes from rail to rail. The output swing to ground is not critical, but it does make things easier to handle, as explained below. Note that the input signals to the comparator are divided in half by R_1 and R_2. This means that we have to provide a ramp signal only to $V_{ref}/2 = 2.5$ V.

If a microcontroller with a 1-μs minimum clock time is used, it will be necessary to count for 4096 μs to provide 12 bits of accuracy. However, there will be some slight variation in component values and in the microcontroller counts (explained below in the software discussion). With this in mind, it is desirable to count to 14 bits and truncate the result to 12 bits. Therefore, we will need approximately 16,000 μs. The exact number of counts is not important since the ratio of two or more measurements will determine the result. All that is important is that there are enough counts under all component tolerance stack-ups (i.e., integrating capacitor) to maintain the desired resolution.

Determining integrator components

Step 1: Determine capacitance. By arbitrarily setting the charge current to about 7.5 μA, the capacitance can be calculated from

$$C = \frac{7.5\ \mu\text{A}\ (16\ \text{ms})}{2.5\ \text{V}} = 0.048\ \mu\text{F}$$

Select $C = 0.047$ μF (film).

Step 2: Determine the bias voltage. It is desirable to have the discharge cycle be much shorter than the charge cycle, to reduce the time between measurements. This can be accomplished by using a bias voltage on the comparator just above ground. It can be arbitrarily set equal to about 0.75 V. With a 5-V logic control swing, the discharge current will then be about (5 − 0.75)/0.75, or 5.7, times faster than the charge cycle. Therefore the divider network is

$$0.75\ \text{V} = \frac{(5\ \text{V})(R_6)}{R_5 + R_6}$$

Let $R_5 = 220$ kΩ, and solving for R_6 yields

$$R_6 = \frac{165\ \text{k}\Omega}{4.25} = 38.8\ \text{k}\Omega$$

Select $R_6 = 38.3$ kΩ.

Step 3: Determine integrator minus input resistance. Since the current is set to 7.5 μA and the delta voltage across the input resistor during a charge cycle is 0.75 V, the input resistor is

$$R_4 = \frac{0.75\ \text{V}}{7.5\ \mu\text{A}} = 100\ \text{k}\Omega$$

Full 0- to 5-V design

Although the integrating amplifier can swing near rail to rail (within about 10 mV), it cannot go all the way down. Even if the output could go all the way to ground, it would have to be stopped prior to this so that the microcontroller would have time to change the direction and start the timer. This is a very important point to understand. If the amplifier is allowed to "bottom out," the accuracy will be severely degraded. In this situation, the input control which is at 5 V will continue to charge the capacitor, thus driving the negative input beyond 0.75 V. This will result in a huge delay at the start of a conversion, thus causing severe linearity error.

For measurements that must include ground, V_z can be set to a voltage sufficiently higher than ground to give the microcontroller sufficient time to control the ramp. In this example it is set to $V_{ref}/65$, or about 77 mV. The actual voltage is not important since the same value is added to each signal, including the ground input measurement.

Circuit operation

For absolute measurements, the reference voltage will have to be measured as well. Each input signal is measured in exactly the same way, so the counts are proportional to the measured input signal levels. First, the ramp is driven down while ground is selected at the multiplexer input. After the "zero" threshold at the comparator is crossed, the ramp is reversed to go in a positive direction. At the moment the threshold is crossed again, the microcontroller timer is started, and then the desired input channel is selected. The timer keeps counting until the new comparator threshold V_{in} is crossed.

The beauty of this technique is that all the hardware errors automatically cancel. This is accomplished by controlling the ramp generator so that the comparator thresholds for ground and the input signals are approached from the same direction. In this example, this is from the low side with the ramp in a positive-going direction. Since the total voltage offset will be the same magnitude and polarity at the start and stop of the timer, the ramp voltage V_r measured will simply be

$$V_r = \frac{R_2}{R_1 + R_2} [V_{in} + V_z + V_{os} - (V_z + V_{os})]$$

$$= V_{in} \frac{R_2}{R_1 + R_2}$$

or if $R_1 = R_2$,

$$V_r = \frac{V_{\text{in}}}{2}$$

Limited-range design

Circuitry following the multiplexer output (shaded box) in Fig. 5.3 is optional for limited input ranges. Amplifier A_3 is a noninverting summer that adds a small voltage V_z to each input. This amplifier is required only if it is necessary to measure the input signals very near ground (i.e., <30 mV). Buffer amplifier A_2 provides high-input-impedance buffering for amplifier A_3. Also, A_2 can prevent unacceptable V_{os} error in the comparator due to mismatches in the input resistances if A_3 is omitted.

Most control applications do not require the input signal to be measured down to ground. As mentioned above, these applications do not require op amp A_3 in Fig. 5.3. To eliminate A_3, we need to select a "zero voltage" above ground for starting the timer. In this example, we chose $V_{\text{ref}}/65$. The reason for this ratio is to simplify computation of the results. Since the entire range is not measured, it is necessary to add the counts missed to each result so that linearity is preserved. Using a resistor ratio of $1/65$ allows for 1 percent resistors to be used since the potential error from this is only $2/65$ percent (worst case). Another benefit is that the mathematics is greatly simplified. Since the measurement actually covers only $64/65$ of full scale, the first $1/65$ is not measured. As will be shown below, we will have to add $1/64$ of the V_{ref} measurement (T_{ref}) to each input.

$$V_z = \frac{V_{\text{ref}}}{65}$$

$T_{\text{ref}} = (V_{\text{ref}})(64/65)$ (This is what's actually measured when V_{ref} is selected.)

Rearranging yields

$$\frac{\text{Counts}}{\text{Volt}} = \frac{T_{\text{ref}}}{(V_{\text{ref}})(64/65)}$$

$$T_z \text{ counts missed} = \frac{\Delta V_z \text{ counts}}{\text{volt}}$$

$$= \frac{1}{65V_{\text{ref}}} \frac{65V_{\text{ref}}T_{\text{ref}}}{64}$$

$$= \frac{T_{\text{ref}}}{64}$$

To determine T_z, simply take the counts for V_{ref} and shift right 6 times (this is the same as dividing by 64), and add this constant to each measurement. This results in a less expensive circuit while still providing 12-bit accuracy for an input range of 77 mV to 5 V.

There is a subtle advantage in using the resistor divider (R_1 and R_2) in the minus comparator input. This allows for a single supply to be used for an input range larger than V_{cc} − 1.5 V and avoids the potential problem of the limited comparator input range. However, when consideration is given to eliminating the buffer prior to the comparator, it is important to determine the potential offset voltages V_{os} that can result from differences in input resistances.

$$V_{\text{os error}} = I_b(\text{delta input resistances})$$

If the offset errors are unacceptable, possibly a comparator with lower input bias current such as the National Semiconductor LP339 can be used. With the LP339 I_b of only 25 nA (maximum), a much wider range of input resistances can be tolerated. Other comparator specifications such as speed and V_{os} are not critical since these conditions are equal for all measurements.

Software Techniques

This section shows a software routine for performing the 12-bit single-slope conversion for either 0 to full scale or a limited-range A/D converter. To complete the software, a division routine will have to be included to take the ratio of the input counts. The technique used in the *RC* A/D converter example in Chap. 4 can be used to accomplish this, except that a 32 × 16 division routine should be used with the lower 16-bit word of the numerator padded with zeros. For applications that are using the limited-range technique, the missed counts (T_z) would have been added to each time measurement prior to the division operation. This technique is explained in detail later in this chapter.

Figure 5.4*a*, is a flowchart describing how to perform the single-slope A/D conversion based on a polling method. The inputs (with G0) are monitored within a tight software loop to determine when to store the timer values. It is very important to minimize the software loop used for polling the comparator output. The number of software cycles within the tight loop will translate to jitter in the measurement. You can think of this as "aperture" uncertainty, and it will show up as noise in the final measurement.

The software routine in Fig. 5.5 is well described, but some of the subtle points need to be explained. As mentioned above, the software loop for monitoring the comparator threshold crossings needs to be very tight. This is easily done by using a powerful feature of the

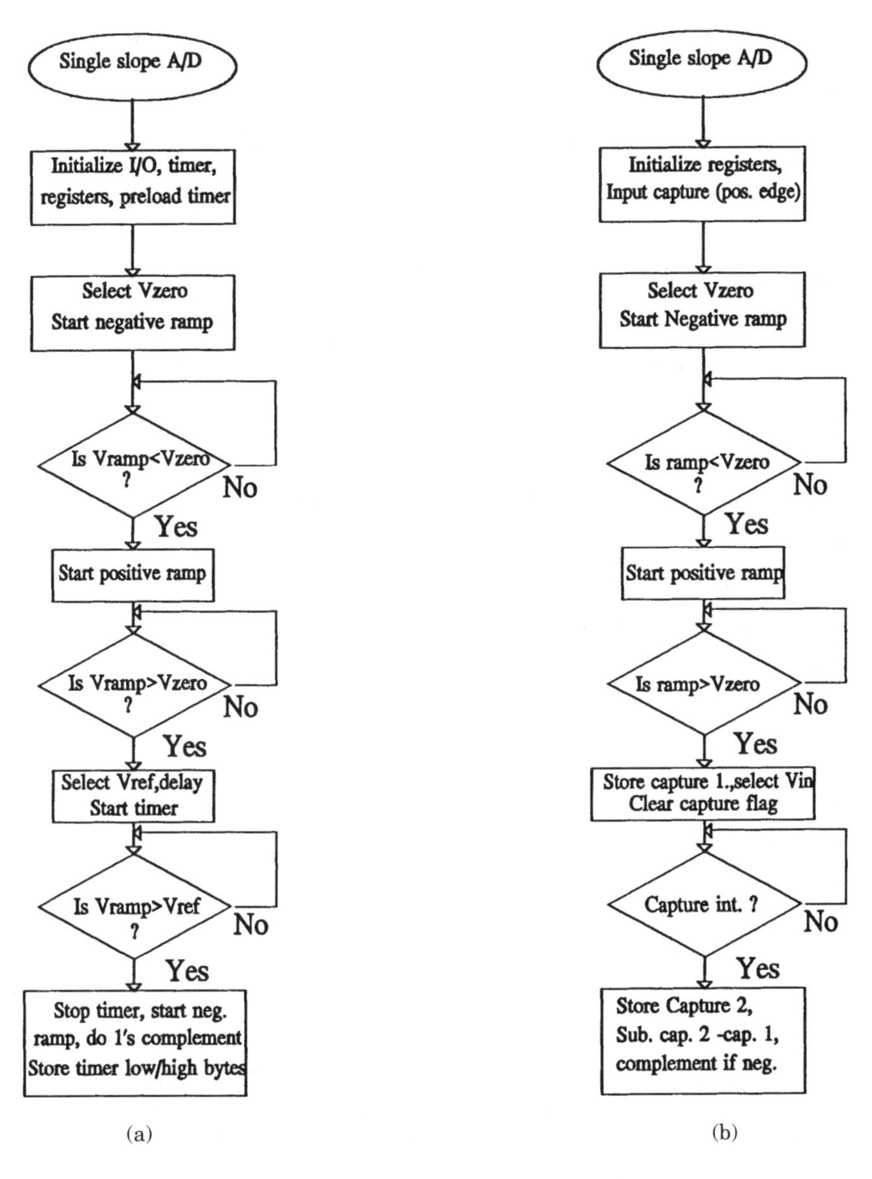

Figure 5.4 Flowcharts for single-slope A/D conversion. (*a*) Polling method; (*b*) Input capture method.

COP800 microcontroller where a [B] pointer register is used for a single-cycle instruction, as described below.

In the code listing, the [B] register is first loaded with the address for the G input register. The multifunction instruction (IFBIT 0,[B]) will then check the comparator output at pin G0 and do the next instruction until G0 is low. Upon the falling edge of G0, the JP POLL instruction will be skipped. This creates a 4-cycle tight polling loop.

```
;THIS IS A SINGLE SLOPE A/D POLLING ROUTINE FOR THE BASIC
;COP800 FAMILY OF MICROCONTROLLERS THAT COUNTS TO 14 B
;FOR 12 BITS OF ACCURACY. Vref IS STORED IN RAM 00,01,
;AND Vin IS STORED IN RAM 02,03.
;
.CHIP 820
TLOW=0EA
THIGH=0EB
LCONF=0D1
LDATA=0D0
GCONF=0D5
GDATA=0D4
GIN=0D6
CNTRL=0EE
;INITIALIZE REGISTERS AND TIMER
        LD LDATA,#01 ;SETUP LPORT
        LD LCONF,#0FF ;SET Lport=OUTPUTS
        LD B,#0D6    ;SET B REG POINTING TO GIN REG
        LD GDATA,#01 ;SET G0= WEAK PULL UP
        LD GCONF,#030 ;G4,5=OUTPUTS; G0,1,2,3,6 =INPUTS
        LD CNTRL,#08A ;SELECT TIMER MODE, EN. MICRO-WIRE
;
;RESET CAP AND COUNT VREF
        LD TLOW,#0FF
        LD THIGH,#0FF
        LD LDATA,#01D  ;SELECT Vzero ON CHANNEL 7
        SBIT 0,LDATA   ;START NEGATIVE RAMP, L0=HIGH
WAIT:   IFBIT 0,[B]    ;WAIT UNTIL COMPARATOR IS HIGH
        JP RAMP
        JP WAIT
;CAP IS RESET
RAMP:   RBIT 0,LDATA   ;START POSITIVE RAMP
LOOP:   IFBIT 0,[B]    ;CHECK FOR + Vzero CROSSING
        JP LOOP
        LD LDATA,#018 ;SELECT Vref ON CHANNEL 6
        NOP           ;PROVIDE EQUAL DELAY
        SBIT 4,CNTRL  ;START TIMER
POLL:   IFBIT 0,[B]   ;WAIT UNTIL RAMP IS > Vref
        JP POLL
        SBIT 0,LDATA  ;START NEGATIVE RAMP
        RBIT 4,CNTRL  ;STOP TIMER
        LD A,TLOW
        XOR A,#0FF    ;INVERT A
        X A,00        ;STORE Vref RESULT IN RAM 00 & 01
        LD A,THIGH
        XOR A,#0FF    ;INVERT A
        X A,01
;
;START Vin MEASUREMENT
        LD TLOW,#0FF
        LD THIGH,#0FF
        LD LDATA,#01D  ;SELECT Vzero ON CHANNEL 7
WAIT1:  IFBIT 0,[B]    ;WAIT UNTIL COMPARATOR IS HIGH
        JP RAMP1
        JP WAIT1
;CAP IS RESET
RAMP1:  RBIT 0,LDATA   ;START POSITIVE RAMP
LOOP1:  IFBIT 0,[B]    ;CHECK FOR + Vzero CROSSING
        JP LOOP1
        LD LDATA,#04  ;SELECT Vin ON CHANNEL 1
        NOP           ;EQUALIZE TIME TO START TIMER
        SBIT 4,CNTRL  ;START TIMER
POLL1:  IFBIT 0,[B]   ;SIT AROUND AND DO NOTHING UNTIL
        JP POLL1      ;RAMP IS > THAN Vref
        NOP           ;EQUALIZE TIME TO STOP TIMER
        NOP
        NOP
        NOP
        RBIT 4,CNTRL ;STOP TIMER
        LD A,TLOW
        XOR A,#0FF    ;INVERT A
        X A,02        ;STORE Vin LOW RESULT IN RAM 02
        LD A,THIGH
        XOR A,#0FF    ;INVERT A
        X A,03        ;STORE Vin HIGH IN RAM 03
.END
```

Figure 5.5 Software routine for polling method.

Tight polling loop

```
cycles
1    POLL:   IFBIT 0,[B]    ;check comparator output
3            JP POLL        ;keep looping until low
```

The accuracy of the polling method is largely determined by the amount of jitter. Although the average counts will balance out, this example counts to 14 bits to reduce the jitter to ± 1 count.

Another very important software technique ensures that the time delays (instruction cycles) for starting and stopping the timer are exactly the same. By keeping the number of instruction cycles equal, there will be no timing error. In this example, when the comparator threshold is crossed with V_z selected, the polling loop is exited, and the following instructions are executed:

```
cycles
3    LD LDATA,#018    ;SELECT Vref
1    NOP
```

```
4       SBIT 4,CNTRL        ;START TIMER
```

The timer will continue to run until the ramp exceeds V_{ref}. At that time, the timer is stopped by using the following sequence:

```
4       SBIT 0,LDATA        ;START NEGATIVE RAMP
4       RBIT 4,CNTRL        ;STOP TIMER
```

Note that there is a delay of 8 instruction cycles after each comparator threshold crossing. Therefore the timer accuracy will be unaffected.

Software using the input capture method

When there is a timer available that can measure the time intervals between input edges (i.e., input capture), the conversions can be performed independent of the software. In contrast with the polling method, the input capture allows the timing to be done automatically once the conversion is under way, thus freeing up the microcontroller for other tasks. The flowchart in Fig. 5.4*b* illustrates the procedure. It is very similar to the polling routine, except in the way the timer is utilized. The timer can be free-running for taking snapshots or captures at positive comparator threshold crossings. An important point to make with this approach is that there could be a false capture when switching from a ground (or V_z) to V_{in}. By resetting the capture flag after V_{in} is selected, the problem is avoided.

Handling of interrupts

It is still possible to use a relatively software-independent technique without using an input capture timer. After the V_z threshold is crossed and V_{in} is selected, the external interrupt can be enabled for a positive edge, and the timer started. This will free up the microcontroller for other time-critical functions. Once the comparator threshold is again crossed, an interrupt will be generated. Depending on how many parameters need to be stored, the total number of clock cycles will have to be added as a delay when the timer is started, or subtracted from each result. This will be a minimum of 7 cycles in the COP800 for pushing the program counter on the stack and incrementing the counter by 2. As long as the delays are the same for both starting and stopping the timer, the delays will automatically cancel. Since the instruction in progress at the time of the interrupt will determine the total interrupt latency, the average instruction time (2 cycles) should be added also.

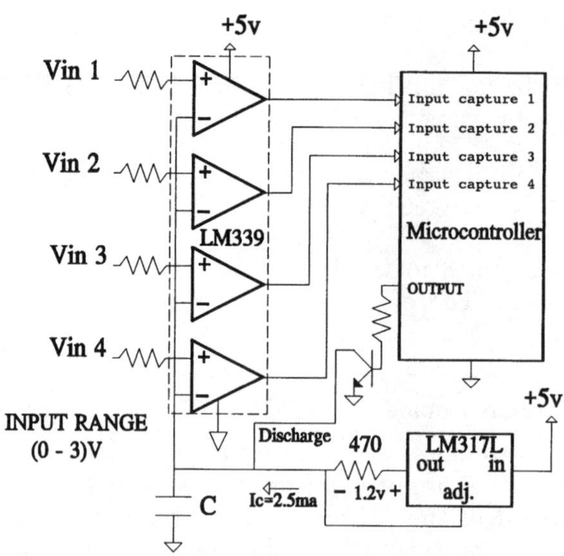

Figure 5.6 Low-cost four-channel single-slope A/D converter.

Simple four-channel design example

For systems that do not require accuracy better than the difference in comparator input offset voltages, a very simple four-channel single-slope circuit can be created. For example, the circuit in Fig. 5.6 uses only an LM339 quad comparator and an LM317L adjustable regulator along with the microcontroller, with the LM317L configured as a constant-current source with a capacitor for providing a ramp signal.

This circuit works by starting with the capacitor discharged by either the output of the microcontroller or an optional transistor (i.e., a 2N7000 low-cost MOSFET or small-signal bipolar). For the microcontroller output to be used, the circuit must be capable of tristating at the start of the conversion. It must also provide a small (<1 LSB) voltage drop above ground while sinking 2.5 mA to minimize the capacitor voltage prior to starting a conversion.

The advantage of this design is that all four channels are measured simultaneously, with each comparator output controlling one of four input capture timers. Thus, four conversions can be performed while the microcontroller is free to work on other tasks. Microcontrollers such as the COP888 series from National Semiconductor can be selected with multiple-input-capture capabilities. If the selected microcontroller does not have a timer with input-capture capabilities, then the feasibility of the polling method should be evaluated, although the polling method may be too slow if all four channels are polled simultaneously. This is due to the large number of instruction cycles to monitor multiple comparator outputs.

Ratiometric measurements allow for the circuit in Fig. 5.6 to measure four unknown inputs. For absolute measurements, only three unknowns can be measured since one channel must be reserved for the reference. The circuit can achieve resolution of up to 10 bits quite easily. The accuracy of this circuit is probably limited, though, to about 8 bits. This is mainly due to each input using a different comparator, thus causing an offset error that cannot be canceled. As before, the unknown inputs can be determined by taking the ratio of the individual timer counts.

To size the capacitor with a charge current of 2.5 mA, it is necessary that the input range, resolution, and clock rate be known. For example, if we are using a 1-μs clock and an input range of 0 to 3 V and if we desire 9 bits of resolution (512 counts), the capacitor can be calculated from

$$I = \frac{C\,dv}{dt}$$

$$C = \frac{(2.5\text{ mA})(512)\mu\text{s}}{3\text{ V}} = 0.42\ \mu\text{F}$$

Select $C = 0.47$ μF.

Note that this design uses only a single + 5-V supply for a usable input range of 0 to 3 V. With this voltage, the usable input range is limited to 0 to 5 − 1.5 V, or 0 to 3.5 V, due to the LM339 comparator and 5 − 2 = 3 V due to the minimum dropout of the LM317L. Keep in mind that when the same supply is used for the digital circuitry, it is very important to adequately filter the power lines for the analog circuitry. For greater input range, the supplies for the analog ICs can be increased.

Dual- or Multislope A/D Converters

Dual- or multislope converters operate on the principle of integrating the unknown input and then comparing the integration times with a reference cycle. The basic approach is to use just two slopes (dual), as shown in Fig. 5.7. Although this is a simple procedure, it suffers from some errors. Among the most severe is the combined input offset errors of the analog circuitry and the capacitor dielectric absorption.

The circuit in Fig. 5.7 operates by first switching in the unknown input signal and then integrating for the full-scale number of counts. Following this cycle, the reference is switched in. Providing the reference is of opposite polarity, the ramp will be driven back toward ground. The time that it takes for the ramp to again reach the com-

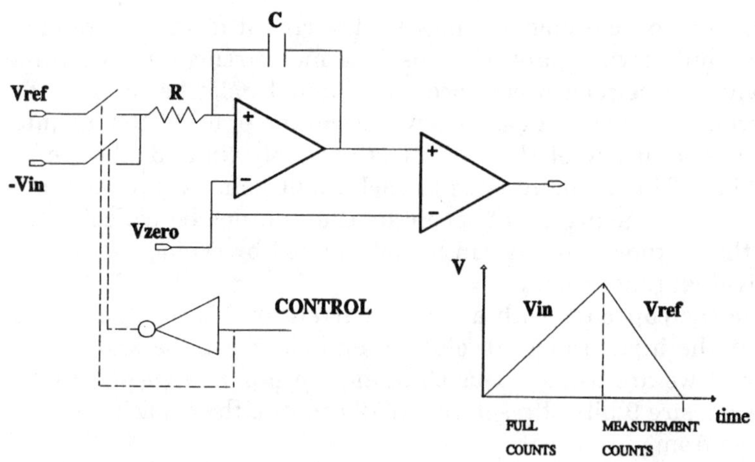

Figure 5.7 Basic dual-slope A/D converter and waveforms.

parator threshold of ground will be directly proportional to the unknown input signal. Since the circuit uses the same *RC* time constant for the integrator, the component tolerances will be the same for both the integration and deintegration cycles. Therefore, these potential errors will cancel. However, there is still a serious problem left to deal with. Since the analog circuitry (i.e., integrator) will have some finite amount of V_{os} that is integrated along with the input signals, the output counts will include this offset. The problem is that a changing slope will cause the V_{os} problem to be additive, not cancel.

Benefits of multislope method

To eliminate V_{os} errors, the multislope method is utilized to achieve the accuracies generally associated with this class of converter (that is, >16 bits).

The benefits of using multiple slopes include

1. Increased range
2. Increased accuracy or resolution
3. Increased speed

Larger input ranges can be measured by forcing a limitation on the maximum integration time. Before the integrator ramp is allowed to exceed the output range, the integrator can be driven in an opposite direction by deintegrating with V_{ref} of opposite polarity. This process can be repeated a number of times, thus expanding the input range proportionally. For example, the integrator can be ramped up by using the input signal for a period of ¼ full counts and then deintegrated with a 5-V reference. If this process is repeated 4 times, the full voltage range will be expanded to 20 V. By keeping track of the total deintegration counts, the full resolution from 0 to 20 V can be achieved with only a + 5-V reference.

Most important, the multislope method will increase the accuracy substantially over that of the basic dual-slope method. This can be accomplished by periodically performing an autozero cycle for cancellation of the total circuit offset-voltage errors. These errors are caused by the circuitry from the input signal up to the integrator. The offset errors will not automatically cancel (as they will with the single-slope converters), and a method must be used to eliminate them.

Offset-voltage correction techniques

Several methods can be used to correct for offset voltages. Some methods require the assistance of the system processor, while other techniques can be considered automatic. Among the most popular options are

1. Quad-slope integration
2. Autozero capacitor
3. Zero integration

Quad-slope integration

The quad-slope integration method was introduced by Analog Devices several years ago, and it works by including two extra integration cycles in addition to the normal V_{in} and V_{ref} of the dual-slope process. It is based on the principle of using two extra integration cycles with voltages of equal magnitude but opposite polarity. As shown in Fig. 5.8, the integrator positive input is biased at $V_{ref}/2$. Using $V_{ref}/2$ makes it possible to provide the integration voltages of $\pm V_{ref}/2$ by switching in V_{ref} and ground at the input.

The trick with this approach is to integrate ground and V_{ref} for a period of $T/2$ each, for a total zero cycle of T. In Fig. 5.9, it is assumed that there is a positive net offset error adding to the input signal integration. The positive offset voltage will cause a longer deintegration

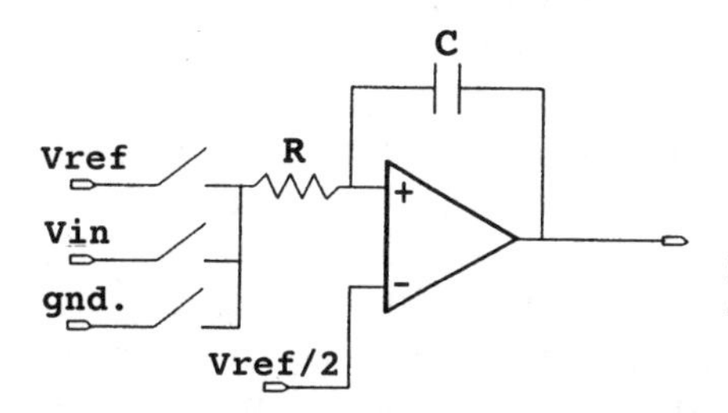

Figure 5.8 Example of quad-slope circuit.

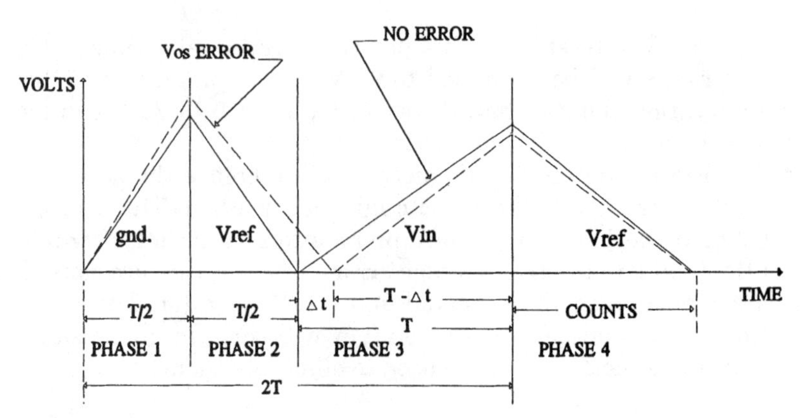

Figure 5.9 Quad-slope waveforms.

time with V_{ref}. This is due to the higher initial ramp while integrating an input at ground. Since the zero cycle should occur for a total of T (full resolution counts), the amount of overcounts delta t (or undercounts if V_{os} is negative) will represent an offset error. This will also be proportional to the V_{os} experienced in the actual measurement since it is also integrated for a period T. The error counts at the end of phase 2 will then cause the main integration of V_{in} to be shortened by the same amount—providing that the integration of V_{in} is terminated after a total conversion time of $2T$ (includes phases 1, 2, and 3). Since the offset counts at the end of phase 2 will steal the same amount of counts away from phase 3, the final result will cancel the offset. The actual technique is a little more complicated than that described above, but the basic idea is the same.

With the above example, the input signals can be plus or minus in polarity and must be less than $V_{ref}/2$ in magnitude. This is to ensure that there will always be a positive integrator ramp.

Autozero capacitor offset correction

Figure 5.10 is an example of a much easier method for performing an automatic offset correction. This figure represents the components necessary to explain how the autozero capacitor technique works. For a more detailed example of this method, the reader is referred to the data sheets on the two-chip solution from Harris Semiconductor (ICL8068, ICL7104). Prior to each conversion, the autozero mode is activated by closing switches S_1 and S_2 while leaving all others open. Zero input voltage is applied to the A_1 positive input, and the total offset of A_1 and A_2 will be integrated by A_2. Note that with S_1 closed, the integrator includes comparator A_3 within its closed loop. In this arrangement, the autozero capacitor (C_z) will be either charged or discharged to force the A_2 positive and negative inputs to equalize. When this occurs, the integrator output will be a flat response since the net

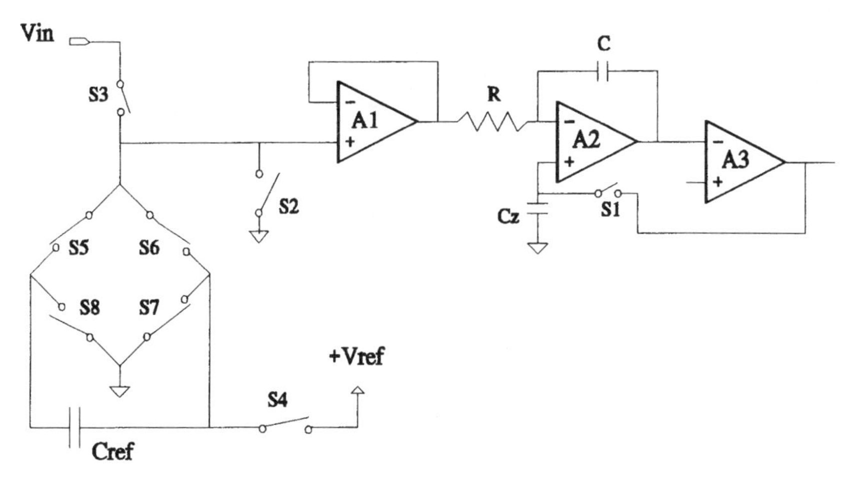

Figure 5.10 Example of an autozero capacitor.

input voltage is zero. Opening S_1 will then cause C_z to hold this offset correction voltage in preparation for starting a conversion.

Also shown in Fig. 5.10 is a set of switches connecting V_{ref} to the input buffer. Remember that in order for the integrator to function properly, the inputs must have opposite polarities with respect to the reference. To make it possible to use a positive reference, the following switch arrangement can be used. Switches S_4 and S_9 are closed (S_6, S_7, and S_8 are open) prior to each deintegration phase, thus charging C_{ref} to V_{ref}. Following this, switches S_7 and S_8 are closed (S_4, S_6, and S_9 are open) to effectively switch in $-V_{ref}$ for integration.

After the autozero cycle is performed, the standard dual-slope measurement can be taken for an output count that effectively takes out the offset errors. There is some limitation, however; some amount of charge injection will be present from S_1 opening, thus causing an error in the offset correction held on C_z. Depending on the type of integrator and the size of the capacitor, the integrator leakage current may also be an error factor.

Zero integration offset correction

The Maxim Semiconductor MAX7129 uses a much different approach for solving the zero offset problem. Instead of requiring a zero capacitor, this technique uses a full conversion sequence with the input buffer shorted to ground. The zero integration counts are then digitally subtracted from the actual result. There are several advantages to this technique. One is the elimination of the charge injection and bias current problems since the autozero capacitor is not required. Thus the accuracy is improved compared to the previously mentioned methods. Figure 5.11 illustrates key portions of the analog circuitry required in

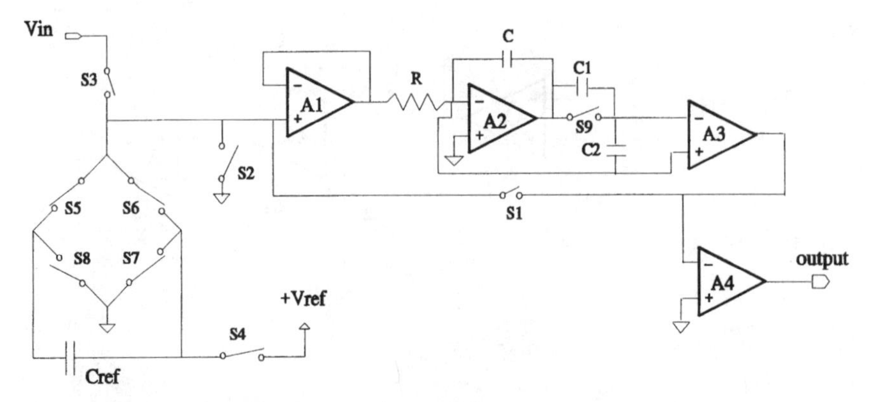

Figure 5.11 Example of a zero-integration circuit.

the MAX7129. The switch arrangement in the front end is similar to the previous example in Fig. 5.10, where V_{ref} is used to charge C_{ref} for voltage inversion. This is where the similarity ends, however.

Referring again to Fig. 5.11, note that there are extra components S_1 and capacitors C_1 and C_2 in the circuit. These elements are used to greatly increase the resolution and speed of the integrating converter. The MAX7129 operates by first performing a 3½-digit conversion. After the input signal is integrated, the deintegration phase is activated by selecting V_{ref}. This will continue until the comparator threshold is crossed.

Shown in the enlarged shaded region of Fig. 5.12, however, is one clock overshoot since the ramp can be stopped only on clock transitions. The overshoot residual voltage is then multiplied by 10, and another full conversion is performed. Following this second integration cycle, a third is performed. With each integration cycle, the residual conversions are multiplied by 10. This is accomplished by switch-

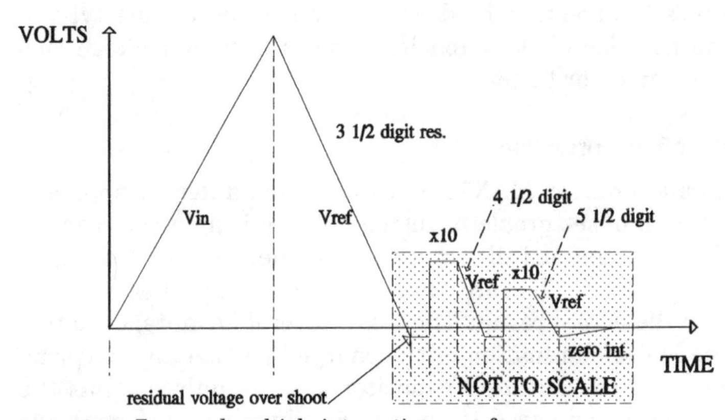

Figure 5.12 Zero- and multiple-integration waveforms.

ing in C_1 and C_2, which differ by a factor of 10, and feeding the multiplied signal back to the buffer input for another integration cycle. Since the newly multiplied residual voltage is integrated for a full cycle, the resolution will also be increased by the same amount each cycle. This effectively generates a 5½-digit result with far fewer counts than in previous methods.

As an example, assume that we are measuring a full-scale input. The first integration cycle will take 2×2000 counts (3½ digits). Since the residual voltages will be quite small, the deintegration cycles will be insignificant and the total number of counts from cycles 2 and 3 will be approximately 2×2000 more counts. Therefore, the final resolution of 5½ digits can be achieved with only 8000 counts instead of $2 \times 20{,}000 = 40{,}000$ counts! This is a major advantage since the only real problem with the integrating A/D converters can be the slow conversion time.

Problems with capacitor dielectric absorption

Regardless of the approach chosen for implementing an integrating A/D converter, there will still be a potential problem. This is the problem caused by the dielectric absorption of the integrating capacitor, and it can show up as linearity and ratiometric errors. Polypropylene capacitors are the usual choice for these applications due to their low dielectric absorption factor. Dielectric absorption is a potential problem in multislope converters because of the changing slopes that cause the capacitor to retain a small charge of opposite polarity. This charge will then dissipate during the next integration slope and will be dependent on the previous and present voltage levels. Therefore, the error will be impossible to correct for. Since single-slope converters do not require a changing slope, there is no problem with dielectric absorption.

Strain Gage Application

One of the most common uses of an integrating A/D converter is for measuring a strain gage as part of a weighing scale. In these applications, it is highly desirable to maintain very high accuracy and, more important, repeatability. In the upcoming design example, many system design techniques will be presented that can make it possible to meet the specifications in a very demanding environment.

To develop an appreciation for the difficulties surrounding a typical scale application, let's start with the strain gage itself. Strain gages use a bridge configuration, as shown in Fig. 5.13. Strain gage sensors generate a small differential output voltage, typically 300 mV full scale.

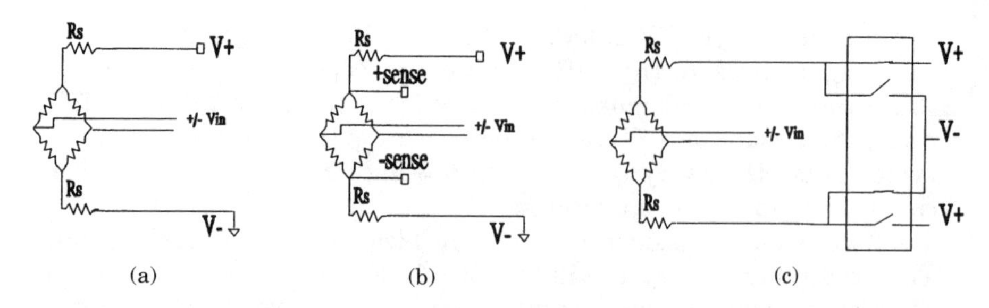

Figure 5.13 Examples of strain gage bridges. (*a*) Basic four-wire; (*b*) six-wire; (*c*) ac excitation.

This output voltage is proportional to the amount of force (weight) placed on the scale and the excitation voltage. Not surprisingly, strain gages are rated for an output at a given excitation voltage. This is because the bridge basically is a resistor divider that produces a small differential voltage centered midway between the excitation voltages.

One of the biggest problems (and there are many) with the use of strain gages is the typical long wires from the sensor to the instrumentation circuitry. Since we are dealing with a very small differential signal, any induced noise could produce gross errors. Obviously, it would be beneficial to increase the input range by increasing the excitation voltage. Unfortunately, this is not possible owing to the linearity problems that would result. With the small differential voltage we are stuck with, it should be no surprise that the signal-to-noise ratio and dc errors from the sensor up to and including the front-end instrumentation amplifier are absolutely critical.

Just to keep things interesting, let's consider the other potential error sources and possible cures.

Potential strain gage error sources:

1. Line pickup of 50 or 60 Hz
2. Leakage current
3. *IR* drops
4. Thermocouple effects of sense wires in circuit board

Potential instrumentation amplifier errors:

1. Input offset voltage V_{os}
2. Limited common-mode rejection ratio (CMRR)

3. Limited open-loop gain (A_{vol})
4. Temperature drift (A_{vol} and V_{os})

Possible cures:

1. Use multiple sense wires for controlling V_{ref} for sensor excitation.
2. Keep thermal gradients low by placing the input sense wires close together and cover if necessary.
3. Use an ac excitation voltage V_{ex} (alternate the polarity of the excitation voltage and compare both positive and negative measurements).
4. Use guard rings around front-end amplifier inputs.
5. Use twisted-pair wires for the sensor and V_{ex}.
6. Shield cabling and attach to the guard ring.

Hardware design example

Now that we have reviewed the basic operation of the strain gage along with its associated challenges, a hardware design example can be examined. Assume that we have this set of requirements:

- 200-mV full-scale input (10-V strain gage)
- 16-bit resolution
- Weighing scale with zero and gain adjustment
- Operating temperature range = 10 to 40°C

Since we are dealing with a very low-level signal from the strain gage sensor (200 mV/65,536 = 3.1 μV per count), the input design will be critical. Referring to Fig. 5.13, note that the input signal from the sensor has a shield around it. At the instrumentation end of the system, there is an amplifier (A_4) that is used for driving the shield at the mean voltage input level. By providing a low impedance drive at very near the input signal levels (that is, 5 V), there will be very little leakage or interference affecting the signal. As a precaution, amplifier A_4 should have an external transistor with short-circuit protection. Not shown in Fig. 5.14, there should also be a guard ring around the input pins to the instrumentation amplifier to further prevent circuit leakage. This should also be connected to the low-impedance output of amplifier A_4.

Note in Fig. 5.14 that there are six wires from the sensor. As mentioned previously, there will likely be some offset voltages caused by

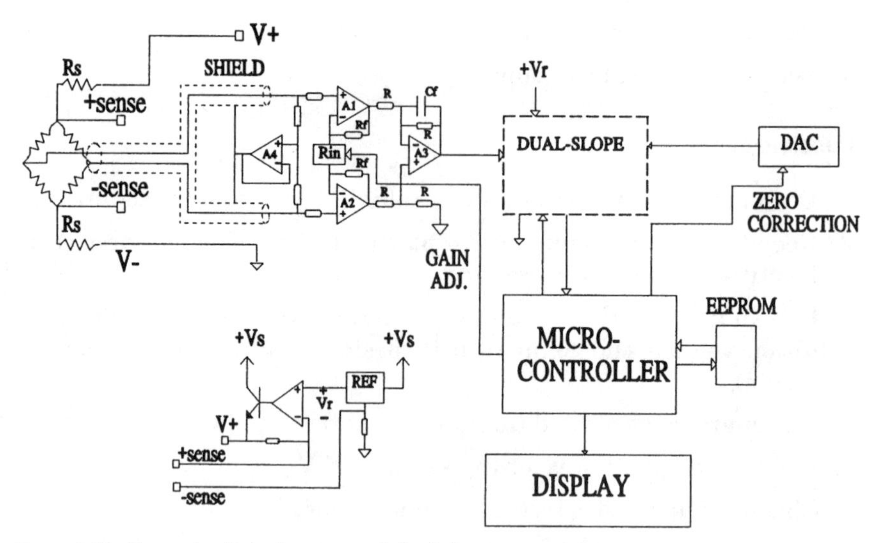

Figure 5.14 Example of strain gage and dual-slope converter.

the resistance in the lead wires powering the sensor. Any changes in the sensor excitation voltage will cause a proportional change in the signal output. Therefore, it is critical that the excitation voltage be stable with respect to the reference used for the A/D converter. By using separate (low-current) sense wires for regulating the voltage at the sensor, this offset voltage can be automatically compensated for.

Note that the same voltage used to drive the sensor is used as the A/D reference. Therefore, this system will operate ratiometrically with any changes in the reference automatically causing the input signal to change proportionally. This being the case, only the reference noise is of concern.

Instrumentation amplifier

The instrumentation amplifier can be either a fully integrated three or four op amp type or a discrete implementation. For this application, a discrete design is chosen. An op amp such as the OP-27 will make a good choice. This op amp typically has an input offset drift of only 0.2 μV/°C and has very low noise at about 4.5 nV/Hz$^{1/2}$. To set the amplifier gain, the input range and reference voltage (full scale) are used.

$$\text{Gain required} = \frac{10\text{ V}}{0.2\text{ V}} = 50$$

By providing gain entirely in the front end with A_1 and A_2, the circuit gain can be described by

$$50 = 1 + \frac{2R_f}{R_{in}}$$

As a means to limit the circuit noise, a simple low-pass filter can be created by placing a capacitor in the differential amplifier (A_3) feedback resistor. Detailed analysis of circuit noise is explained in Chap. 9. With the input being nearly a dc signal, the 3-dB frequency can be set fairly low (that is, 30 Hz):

$$f(3\ \text{dB}) = 30\ \text{Hz} = \frac{1}{6.28\ RC}$$

Gain correction

Since neither the sensor nor the reference will be exact, some full-scale gain adjustment will be required. To accomplish this, a set of resistors in parallel are controlled with an analog switch for setting R_{in}. Use of a precision fixed resistor along with a resistance bank of much larger resistors will provide a fine adjustment. If a course adjustment is required, then an additional second bank of resistors should be used.

Resistor network

The instrumentation circuit resistances need to have very high ratiometric tracking. This is true for the two feedback resistors (R_f) in the front-end gain circuit and for the differential amplifier (A_3) resistors. Any mismatch in these resistors will cause common-mode error. It is important for not only the initial values to be the same, but also the temperature drift. The slight initial tolerances can be calibrated out, but any drifting in values afterward will likely be undesirable. Precision thin-film resistor networks such as those from Electro-Films Inc. are available with accuracies well within the requirements (i.e., up to 0.5 ppm/°C, ratio tolerance of 0.003 percent). This is where the discrete instrumentation amplifier will have an advantage over a fully integrated version. The temperature tracking of the thin-film networks is considerably better than that of the resistors in the integrated solution, making it easy to use. Most precision manufacturers will provide custom thin-film networks without significant engineering cost. To select a resistor network, it is necessary to first determine the required temperature drift:

$$\text{Drift} = \frac{1}{(30°\text{C})(65{,}536)}$$
$$= 5.1 \text{ ppm/°C}$$

For the differential amplifier (A_3), low-value resistors for R (that is, 10 kΩ) are preferable to limit noise. To produce a gain of 50 in the front end, it is desirable to make the two feedback resistors (R_f) reasonably higher. This is so that the analog switch resistance temperature drift as part of R_{in} will be minimized.

Dual-slope circuit

The dual-slope function can be implemented in one of the previously described approaches. Regardless of the approach, this circuit will consist of an integrator, a comparator, and a series of analog switches. Zero scale adjustment can be accomplished either by adjusting the reference for the integrator or by using an amplifier for adding a small dc offset to the inputs. This can be done in the same way as the gain adjustment (i.e., parallel resistors with an analog switch network).

Microcontroller role

The main function of the microcontroller is to provide timing and control of the dual-slope process. In addition to these tasks, the microcontroller must provide operator interface for calibration, interface with the keypad, and display. Calibration is usually performed by first zeroing the scale. With zero weight on the scale, the microcontroller can control the zero resistor bank until the display indicates a zero readout. Next, full scale is adjusted by controlling the gain resistor network. Once the scale is fully calibrated, the analog switch configurations are stored in EEPROM.

As an option, the input measurement counts for zero and maximum weight can be stored in EEPROM, and a mathematical routine used to adjust each measurement. Stated differently, instead of using hardware to adjust the circuit range, a mathematical routine is used.

Chapter

6

Sampling (Successive-Approximation) A/D Converters

Sampling A/D converters are also referred to as *successive-approximation* analog-to-digital converters. The term *successive approximation* is used due to the process of successively testing each bit of resolution from the highest (most significant bit) to the lowest (least significant bit). These types of analog-to-digital converters are by far the most popular today. There are wide ranges in performances and levels of integration to choose from to fit a variety of applications. Earlier, "nonsampling" successive-approximation converters sampled the input as each bit was tested. For example, an 8-bit converter would sample the input 8 times during the conversion process.

Modern, "sampled" successive-approximation converters sample the input only once per conversion. The sample time necessary for the input signal to fully charge the capacitor input circuit (S/H) is the *acquisition time*. This is an optimum time based on the internal resistance and capacitance. However, in an actual application, the acquisition time will largely depend on the external input circuitry. Sampled converters have the advantage over the nonsampled architectures of being able to tolerate input signals changing between bit tests. Conversely, nonsampled converter performance would be degraded if the input signal changed more than ½ LSB before the conversion was completed. This is because the inputs are sampled n times for every conversion, where n is the bits of resolution.

Operation of Sampling A/D Converter

The basic principle behind the sampling A/D converter is to use a digital-to-analog converter (DAC) approximation of the input and make a comparison with the input for each bit of resolution. Following the input signal acquisition, the most significant bit (MSB) is tested first. This is achieved by generating $\frac{1}{2}V_{ref}$ with the DAC and comparing it to the sampled input signal. Shown in Fig. 6.1, the successive-approximation register (SAR) drives the DAC to produce estimates of the input signal. The process is started with the MSB and continues to the least significant bit (LSB), with each estimation more accurately homing in on the input level.

For each bit test, the comparator output will determine if the estimate should stay as a 1 or 0 in the result register. If the comparator indicates that the estimated value is under the input level, then the bit stays set. Otherwise, the bit is reset in the result register.

Sampled-data comparator

Nearly all sampling converters today are built with CMOS technology and use capacitor-based instead of resistor-based ladder networks for the DAC. Ladder networks with capacitors can be easily trimmed for producing a weighted capacitance reference (size directly proportional to value). The sampled-comparator shown in Fig. 6.2 also takes advantage of this architecture. Actually, the circuit in Fig. 6.2 does not really use a comparator; a common inverter is used instead. However, the

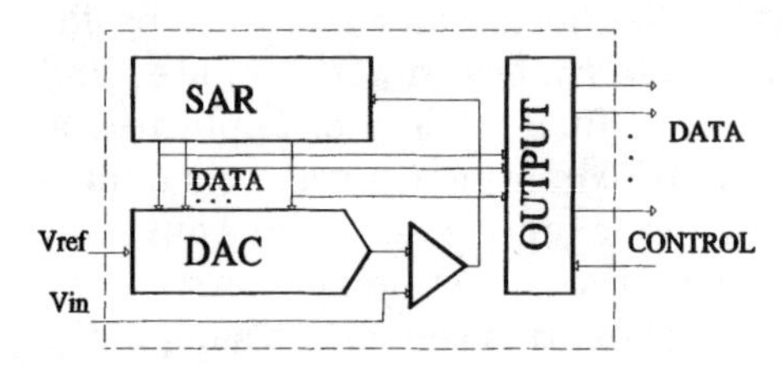

Figure 6.1 Block diagram of typical sampling ADC.

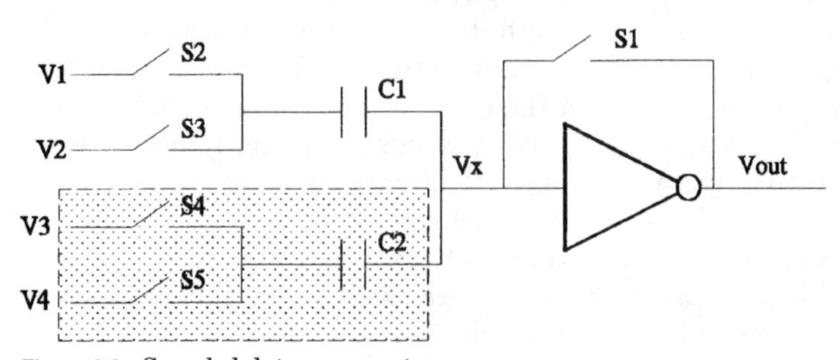

Figure 6.2 Sampled-data comparator.

inverter functions as a comparator when the analog switches are used along with various capacitors. This works by starting with a zero phase (S_1 and S_3 closed, S_2 open). By closing S_1, the inverter will settle to about half the supply level as a result of feedback. This causes the input capacitor C_1 to charge to $V_x - V_2$. Note that this circuit is not susceptible to comparator offset voltage, which is a major advantage over the conventional comparator. Following the zero phase, the compare phase is activated by closing S_2 and opening switches S_1 and S_3.

In the compare phase, the input signal V_1 will cause the inverter input voltage V_x to shift if V_1 is different from V_2. Since V_x was sitting at the threshold of the inverter, any change in V_x will cause the inverter to go either high or low. Thus, a comparison can be made between V_1 and V_2.

Shown in the shaded region of Fig. 6.2 are two more switched inputs. This illustrates a major benefit of this capacitor-based architecture and makes differential measurements quite easy when weighted capacitors are used. By using scaled capacitor values tied to a reference voltage, each bit of the successive-approximation process can be tested. For example, if C_2 is $\frac{1}{4}C_1$, then the effective differential voltage at V_x is

$$V_x = V_1 - V_2 + \frac{V_3 - V_4}{4}$$

Capacitor-based DAC

In Fig. 6.3, the binary weighted capacitors are used to create the DAC. This differs from earlier devices that relied on trimmed resistances in an *R*-2*R* ladder. With S_1 closed, C_7 connected to the input voltage, and all other capacitors tied to ground, a track-and-hold is effectively created. This is inherent in the design and is a major reason why this technique is used. During this autozero mode, C_7 will be

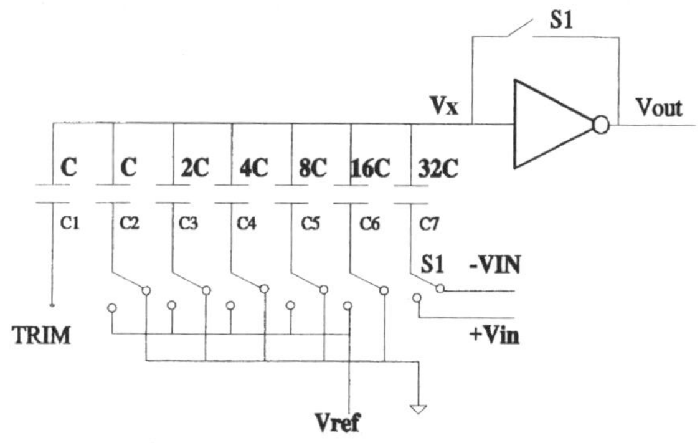

Figure 6.3 Capacitor-based DAC.

forced to charge to $V_{in} - V_x$, and the remaining capacitors to V_x. Note that C_7 is of the same magnitude as the accumulated values of C_1 through C_6. Since the total charge between two capacitors must remain the same, the capacitor voltages will balance out to an equilibrium voltage. This voltage is based on the weighted capacitance and the differential input voltages.

When S_1 is opened, the input voltage V_{in} will be held on the DAC capacitor array, which effectively is a sample-and-hold. The MSB is tested first by switching C_7 from V_{in} to ground and C_6 to V_{ref}. With C_7 equal to one-half the total capacitance, this test will determine if the input voltage is less or greater than $\frac{1}{2}V_{ref}$. The conversions are continued by individually switching in the other capacitors in the order of lesser weighting factor. Finally, the last bit of resolution is obtained by using a combination of capacitors and possibly a small resistor ladder for fine trimming.

Signal Conditioning

Although CMOS sampling A/D converters can differ, there are several similarities. In some types of designs, there may be an input buffer with a sample-and-hold (S/H) for providing a course measurement (that is, one-half full scale). This will speed up the conversion process for charging the input capacitor with the largest step change. Most input stages can be approximated, though, as an analog switch with a series resistance and a holding capacitor. The basic circuit shown in Fig. 6.4 is useful to point out why problems can arise.

External resistance limitations

If the external input resistance R_s is very large, the input capacitance may not have enough time to fully charge before the conversion is started. This will cause undesirable errors that will have to be remedied. As an example, with the following condition, let's determine the maximum external resistance R_s:

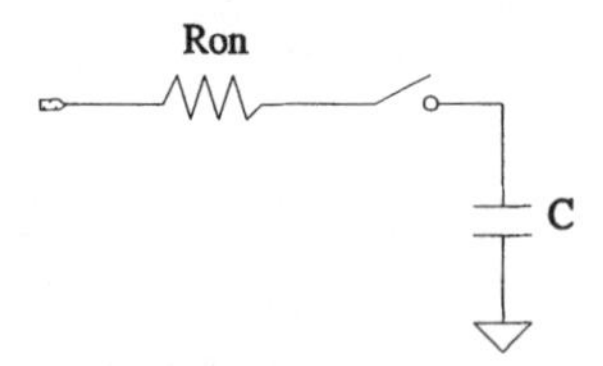

Figure 6.4 Approximation of sampling ADC input.

Circuit conditions:

1. Switch is closed for 2 μs.
2. $R_{on} = 1\ \text{k}\Omega$.
3. $C_{in} = 20\ \text{pF}$.
4. Resolution = 10 bits.

For 10 bits, we will need 8.3 time constants to limit the error below 0.25 LSB.

$$\frac{V_c}{V_{in}} = \frac{1}{1024 \times 4} = e^{-t/RC}$$

Taking the natural logarithm of both sides yields

$$t = 8.3RC \qquad \text{or} \qquad R = \frac{2\ \mu\text{s}}{(8.3)(20\ \text{pF})} = 12\ \text{k}\Omega$$

Therefore,

$$R_s = 12\ \text{k}\Omega - 1\ \text{k}\Omega = 11\ \text{k}\Omega$$

In the above calculation, the maximum series resistance was determined. If a larger value is used, there could be errors produced depending on the input voltage swings. This will certainly be the case for multiple channels if two adjacent channels differ by nearly the full-scale range. What will happen in this case is that the input capacitance will not be fully charged or discharged at the end of the acquisition time. The final voltage held on the input capacitance is then determined by the difference between the previous voltage and the next channel voltage. This is sort of like crosstalk error, since the adjacent channels can affect the selected channel measurement.

To eliminate the problem with high input resistance, several options exist (assuming that R_s cannot be changed). First, the acquisition time can be changed by either altering the converter clock or programming the acquisition time (if possible), although care must be taken to check the converter minimum clock frequency so that internal capacitance leakage does not become a problem. Other options are to reduce the input voltages, select channels in order of voltage magnitude, and perform multiple measurements on a selected channel that differs significantly from a previous measurement.

Input filter pitfalls

When a signal is measured in the presence of noise, it is necessary to attenuate signals higher than one-half the sampling frequency (or higher than the signal of interest) with a low-pass filter. The simplest

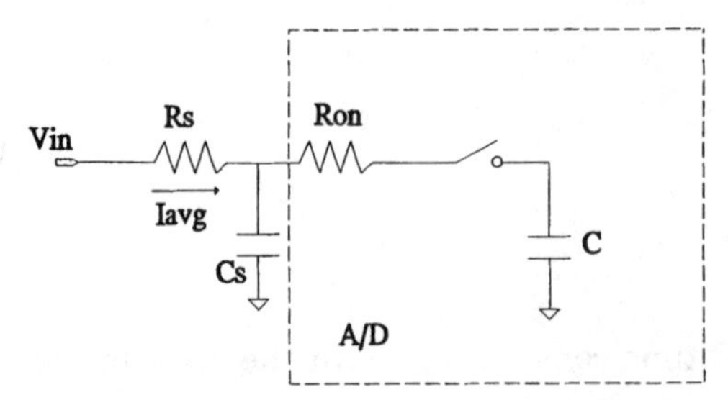

Figure 6.5 Sampling A/D switched input current.

filter is created with an additional input capacitor C_s, as illustrated in Fig. 6.5. Let's assume (1) that the input capacitance C_s is much larger than C and (2) that C is discharged through some internal path. Then each time the switch is closed, current will flow from the input to charge C. This is where a problem can occur. The current seen at the input of the ADC will no longer be short-duration spikes that can be ignored, but will be an average current flowing from C_s. This average current input to the ADC can be approximated by

$$I_{avg} = C_x V_{in} (f)$$

To illustrate the potential problem, assume that the sampling rate is 100 kHz, V_{in} = 5 V, C = 20 pF, and R_s = 20 kΩ. Then the input will have an offset of

$$I_{avg} = 20 \text{ pF } (2.5)\,(100 \text{ k}\Omega) \qquad \text{(capacitor typically tied to } V_{ref}/2)$$

$$= 5\ \mu\text{A}$$

$$V_{os\ error} = I_{avg}\,(20 \text{ k}\Omega) = 5\ \mu\text{A}\,(20 \text{ k}\Omega) = 100 \text{ mV}$$

In the above example, the input capacitance solved one problem, but created another. The 100-mV offset is likely to be much too high and will have to be reduced. To correct this, an amplifier will most likely have to be used for isolation and filtering.

Input amplifier (buffer and filter)

In many circumstances, the input signal will require an input amplifier to scale the analog signal to match the ADC input range. The use of an amplifier can have many advantages. Besides providing gain, the amplifier will reduce the input resistance seen by the ADC, and avoid the potential problems mentioned above. Another benefit of using the amplifier is that an active low-pass filter can be designed. This avoids the ADC switch currents by isolating the filter capacitor from the ADC input. The example in Fig. 6.6 shows the National

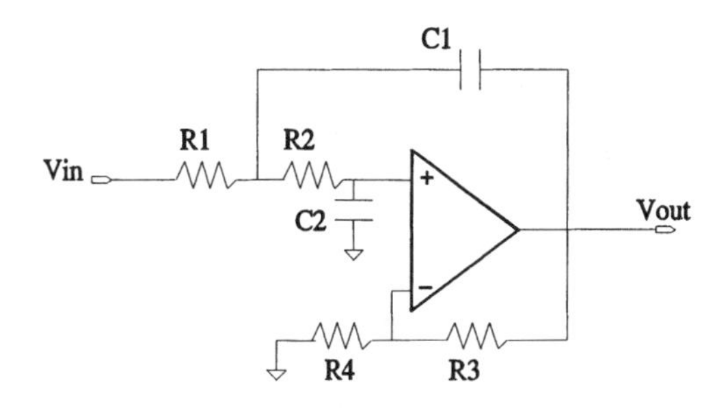

Figure 6.6 Buffer amplifier and filter example.

Semiconductor LMC662 CMOS amplifier. The advantage of the LMC662 is that it can operate from a single supply and can drive the outputs from rail to rail. Therefore the same supply can be used for the amplifier and the ADC. Circuit gain is achieved by setting the ratio of R_3/R_4 as

$$\text{Gain} = 1 + \frac{R_3}{R_4}$$

By selecting proper values for the circuit R's and C's, any of the classical second-order filters can be created. For best performance, the Butterworth-type filter should be chosen. This will provide a smooth roll-off without peaking. One way to design the filter is to use a technique sometimes called *impedance scaling*. By using values from a table for the dampening factor, and frequency-scaled capacitors C_1'' and C_2'', the circuit component values can be found. For the Butterworth filter, $C_1'' = 1.414$, $C_2'' = 0.7071$, and the dampening factor = 0.7072. The following procedure shows how to calculate the values:

Step 1. Calculate C_1' *and* C_2'.

$$C_1' = \frac{C_2''}{6.28\, f_{cp}} \qquad C_2'' = \frac{C_2''}{6.28\, f_{cp}}$$

where f_{cp} is the desired corner frequency with a pole.

Step 2. Select $R_1 = R_2 = R$.

Step 3. Determine C_1 and C_2.

$$C_1 = \frac{C_1'}{R} \qquad C_2 = \frac{C_2'}{R}$$

Note: Choose R so that C_1 (largest value) is a standard value.

Reference Input Requirements

When it is not provided internal to the ADC or when the performance is not adequate, an external reference will have to be applied. Adding a reference is an important task since any errors introduced will directly affect circuit performance. It is important to understand the requirements of the ADC so that performance is not limited. Typically, CMOS sampling ADCs can operate with a reference up to the supply voltage and down to very nearly ground. The data sheet of the selected ADC should be consulted, however. With some sampling ADCs, the zero and linearity errors will get out of hand when V_{ref} is too low.

Regardless of the selected ADC, it is important that the reference voltage be derived from a very low-impedance and low-noise source. Most A/D reference input circuits have low impedance in the range of 1 kΩ. With the low impedance, an *RC* filter is probably not an option to limit noise since this would cause a resistor-divider effect. To solve this problem, there are several standard references that have adequate drive with low output resistance. With this solution, reference filtering is done with a buffer amplifier for isolation.

Some applications may require a low-cost alternative to an integrated reference. The only other option is to use a zener-diode-based discrete solution. Keep in mind that when a low-cost reference such as a zener diode is chosen, the input impedance needs to be low if a filter is added. The circuit in Fig. 6.7 uses an amplifier as a buffer so that the zener noise can be filtered. If high accuracy is desired, a temperature-compensated zener should be used, such as the 1N825A, with a regulated supply setting the zener current. This will keep the voltage stable (matching the data sheet specifications). If necessary, calibration can be done with an optional fine adjustment (potentiometer R_4), as shown in Fig. 6.7. By keeping R_4 much less than R_3, the relatively high-potentiometer-temperature drift will not be an error factor.

Input Protection

When it is possible for the inputs of sampling ADCs to exceed the supply levels, some protection is necessary. This can cause damage to the

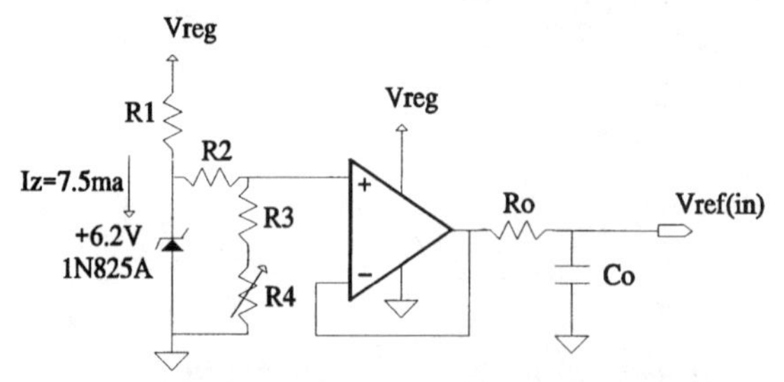

Figure 6.7 Example of a reference input circuit.

device by destroying the protection diodes and ultimately the electrostatic discharge (ESD) circuit. Generally, manufacturers recommend that the ADC input current be limited to about 5 mA, although this is usually a very conservative number that can likely be exceeded.

Overdriving the inputs beyond the supply limits will also cause performance to degrade. Performance can be affected by the overcharging of the input capacitor, thus causing an increased delay to acquire the next signal. Another performance problem can be due to some of the input current from the overdriven channel affecting the selected input signal.

There are a few ways to protect the inputs of the sampling ADC. One of the most commonly used circuits is shown in Fig. 6.8. This circuit uses two diodes for clamping the input signal to within plus or minus one diode drop from the supply levels. Resistor R_1 limits the current through diodes D_1 and D_2. With a reduced input overdrive of $\pm V_d$, resistor R_2 can effectively limit the current into the ADC.

$$I_{\text{in (max)}} = \frac{0.6\ V}{R_2}$$

Another option for limiting the input voltage is to use an op amp that is powered from the same supply as the ADC. An example is the LMC66x-series rail-to-rail output amplifiers. Since the op amp cannot drive past its power supplies, it is impossible for the ADC to be overdriven.

The same protection can be achieved from an amplifier powered by a higher supply than the ADC by merely using a resistor divider in the feedback (Fig. 6.9). By scaling the ratio of the divider properly, the amplifier output voltage V_{in}' to the ADC will be limited.

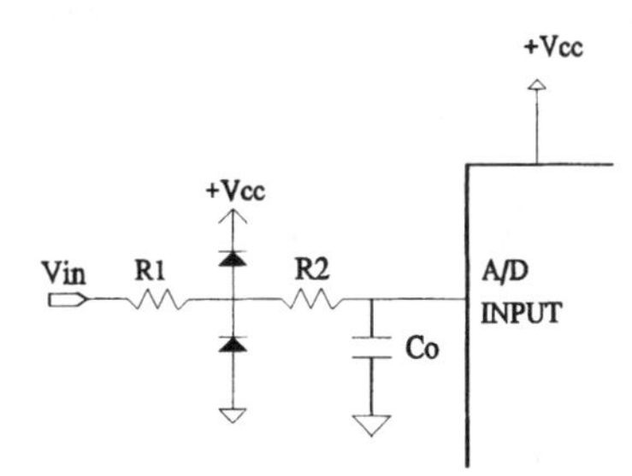

Figure 6.8 Simple input protection circuit.

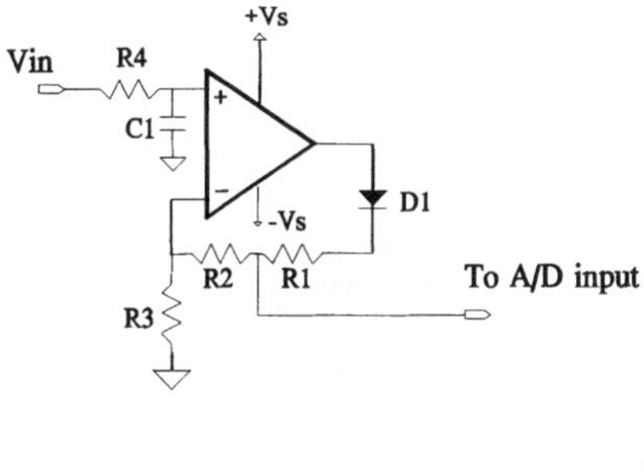

Figure 6.9 Amplifier output limit circuit.

Remember to use care in selecting the resistor values so that the ADC switching currents will not be a problem. The diode in the amplifier output (D_1) is used to prevent the amplifier from driving the ADC input negative in the event of a transient. With an amplifier also providing buffering, a low-pass filter (C_1 and R_4) can be used in the input stage.

Dual-Polarity Measurements With Single Supply

Sometimes it is desirable to make measurements of both positive and negative signals and save the cost of an additional supply. This is possible with the circuit shown in Fig. 6.10.

Input resistors R_1 to R_4 in Fig. 6.10 should be part of a thin-film resistor network so the values will match and track over time versus temperature. This provides a low-cost solution since the values do not have to be exact, only the same. For-high resolution designs, there is little concern about the component tolerances. However, if high accuracy is also desired, the tolerance matching of the input resistors and the V_{os} of the amplifier will have to be selected carefully.

With the network shown in Fig. 6.10, the circuit will measure inputs from $-2V_{ref}$ to $+2V_{ref}$. You can verify this by using the superposition method of alternately short-circuiting V_{ref} and V_{in} and adding the voltages produced at V_x. Then set V_x to the limits of the ADC (i.e., ground and V_{ref}), and solve for V_{in}. Note that R_3 and R_4 are in parallel. This is probably the easiest way to create a stable ratio of $R/2$. Other input ranges can be accommodated by changing the values of resistors R_1 through R_4.

More than likely, the input resistance of the network will have to be higher than the ADC would like to see, to prevent input signal loading. However, remember that the input to the ADC must be driven by a low-impedance source to avoid problems with the input capacitance. This is why there is an input buffer amplifier for driving the

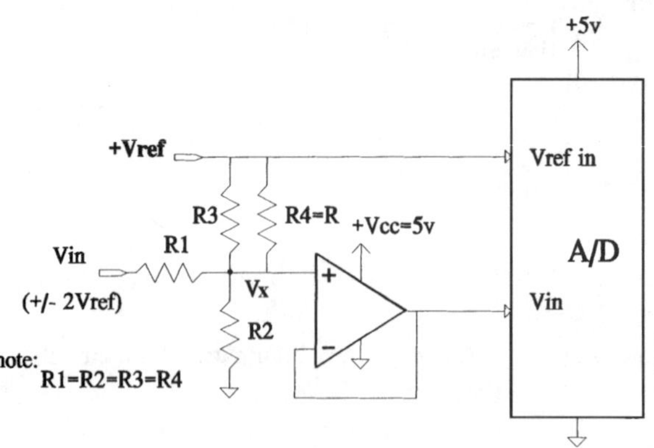

Figure 6.10 Dual-polarity measurement with single supply.

ADC input. Although it is not mandatory, it will provide higher performance. Optimum performance can be achieved with an amplifier such as the National Semiconductor LMC6482 with rail-to-rail input/output and extremely high input impedance. However, it's the rail-to-rail output swing that is most important.

Dithering for Resolution Improvement

Microcontroller designs are usually limited to 8-bit resolution with the on-board ADC. If this is inadequate and the measured signal is near direct current, there is a potential solution. For instance, instead of requiring an external higher-resolution converter, a lower-cost approach may be possible. This will require that some components be added to the microcontroller for using a technique called *dithering*.

Basically, dithering forces the input signal to move between levels of the ADC by adding a small ramp signal to the input. By taking several measurements and averaging the accumulated result, the effective resolution is significantly enhanced. You can think of this in the same way as measuring a signal that has a small amount of noise riding on it. If 50 percent of the noise is above and 50 percent below the signal level, and if the amplitude is greater than 1 LSB, then the dc average is an accurate representation of the true signal. The circuit drawing in Fig. 6.11 shows the basic idea behind how dithering is performed.

The main function of the dithering circuit operates by creating a ramp with amplifiers A_1 and A_2. By sizing $R_1 >> R_2$, the peak-to-peak voltage swing will be limited to a fraction of the analog supply voltages. However, the desired $V_{p\text{-}p}$ signal should be sufficiently larger than 1 LSB to ensure that the ADC thresholds are exceeded. It is also required that 50 percent of the ADC samples be consistently taken

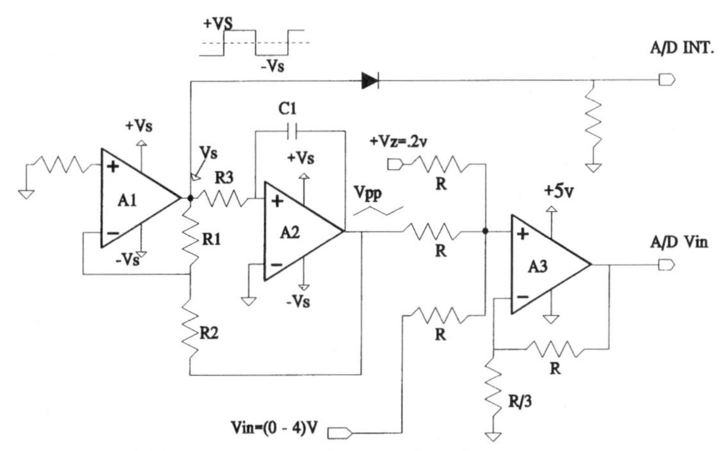

Figure 6.11 Dithering circuit for increased resolution.

from below and 50 percent from above the input signal level. For example, using 10 to 20 times the original A/D LSB should be adequate for determining $V_{p\text{-}p}$, the peak-to-peak voltage. This allows for some slight microcontroller, or external circuit timing, differences. The example below will clarify how to size the ramp $V_{p\text{-}p}$ signal.

Let

$$V_{ref} = 5\text{ V} \qquad \text{and} \qquad \text{initial resolution} = 8\text{ bits}$$

where

$$1\text{ LSB} = \frac{V_{ref}}{256} = 20\text{ mV}$$

$$V_{p\text{-}p} = 10\text{ to }20\text{ times }1\text{ LSB}$$

$$= 200\text{ to }400\text{ mV}$$

To produce the above $V_{p\text{-}p}$ signal, it is necessary to size R_1 and R_2. Amplifier A_1 output swings also play a role in determining $V_{p\text{-}p}$. As an example, let's compute the values of R_1 and R_2 for a 400-mV $V_{p\text{-}p}$. Let $R_1 = 100\text{ k}\Omega$, and determine R_2.

$$\frac{V_{p\text{-}p}}{2} = 0.2\text{ V} = (5\text{ V})\left(\frac{R_2}{R_1}\right)$$

$$R_2 = \frac{(100\text{ k}\Omega)(0.2)}{5} = 4\text{ k}\Omega$$

The small ramp voltage centered on ground is added to the input signal by using the noninverting summer A_3. This requires that A_1 and A_2 be powered from a dual supply. In addition, a small offset voltage V_z equal to $V_{p\text{-}p}/2$ is added to the input signal. Adding V_z prevents the ramp voltage from bottoming out the measurements near ground. In other words, since it is necessary to have the ramp voltage swing above and below the input signal, an input less than $V_{p\text{-}p}/2$ would be a problem.

Besides setting $V_{p\text{-}p}$ levels, the ramp duration must be set. The integrator components R_3 and C_1 and the supply voltages on A_1 are used to determine the ramp duration. It is important to know the period of a complete ramp cycle so that the full number of conversions is made. Remember that we are using several measurements taken above and below the original input signal. This is necessary to preserve the accuracy of the averaging process. Given the number of measurements n desired to be averaged, $V_{p\text{-}p}$, supply voltage swings V_s, and the A/D conversion time, the total ramp duration can be set from

$$dt = (\text{conversion time}) \times n$$

$$V_{\text{p-p}} = \frac{V_s\,dt}{RC}$$

Solve for RC.

Note in Fig. 6.11 that the ramp generator is powered from dual supplies for producing a ramp centered on ground. The supplies for A_1 and A_2 do not have to be precise due to the way the measurements are made. Amplifier A_3 only needs to work from ground to $+5\text{ V} - V_{\text{p-p}}$; therefore it requires only a single supply. This also has the advantage of protecting the ADC input since A_3 will never swing outside the ADC supply.

For applications that require only increased resolution, there is probably no need to trim the circuit. If accuracy is a problem due to combined offset voltages, then the zero voltage V_z can be adjusted. One way to do this is to input 0 V and adjust V_z until a zero result is achieved (i.e., LED output when zero).

The procedure for performing dithering starts with the synchronized pulse from A_1 causing an interrupt. Once the measurements have started, the interrupt edge is changed so that the duration of the ramp can be measured. By measuring the duration of the ramp, ADC measurements can be made at equal increments of time. This allows for component shifts not affecting the accuracy. It is important to take ADC measurements at equal time slots to preserve the resolution. Otherwise, the measurements will appear to be random. The number of measurements needed before an average can be computed will depend on the increased resolution and the desired stability. As the desired resolution is increased, so will the total number of counts. For example, a 12-bit resolution may require approximately 128 measurements. This indicates that the input signal bandwidth must be near direct current!

Interfacing to A/D Converters

There are many options for interfacing between the ADC and the system processor. Basically, from a hardware standpoint, this can be done in a synchronous serial or parallel form. Serial communication usually lends itself to interfacing to a microcontroller or a digital signal processor (DSP). The benefits to using serial communication for the microcontroller are probably obvious: It saves on the required I/O. Most microcontrollers also have built-in synchronous serial hardware to make the communication rather simple. Microprocessors, on the other hand, connect more easily to an ADC with a parallel interface. This is because of the built-in capability of accessing external parallel memory.

Serial interface

Microcontrollers from National Semiconductor, Motorola, and Texas Instruments contain dedicated serial shift registers with special software for operation. These features make interfacing with serial ADCs very simple. For example, with the National Semiconductor COP800 series, first the serial I/O 8-bit register is loaded with the start bit, control bit, and multiplex address. Then conversions are started by pulling the CS line low and holding it low. At the same time, the SIO busy bit is set; this automatically starts the clock for shifting out the microcontroller data into the ADC. Simultaneously, as the data are shifted out, the input data are clocked in the SIO register. Depending on the number of channels, etc., the first byte will have to be padded with 0s. This makes the data communication process much easier since the SIO shifting will automatically stop after 8 bits and will reset the busy flag. Referring to the timing diagram of a typical serial ADC (that is, ADC0838) in Fig. 6.12, the MSB will be shifted out after one-half clock delay. Note that there is a leading zero that must be ignored in the result. The actual result is then shifted in the SIO register by setting the busy bit again to generate the clock again. Depending on the SE pin level, the data can also be shifted out, starting with the LSB first.

Figure 6.13 shows a typical serial interface to a microcontroller. Serial communication to the microcontroller is compatible with either National Semiconductor (Microwire) or Motorola (SPI) serial peripheral interface. These two standards are nearly identical, with the only difference being in the phase of the clock when the data are sampled.

Note that there are four solid lines. These lines are the minimum connections to the microcontroller for accessing the ADC. The separate conversion clock (dotted line) is sometimes provided by the ADC for controlling the conversion and acquisition speed. With a separate conversion and serial shift clock, it is possible to optimize the conver-

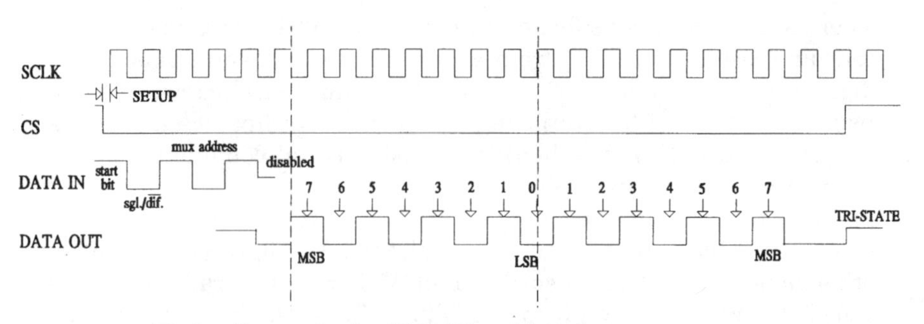

Figure 6.12 Timing diagram for the ADC0838 serial ADC.

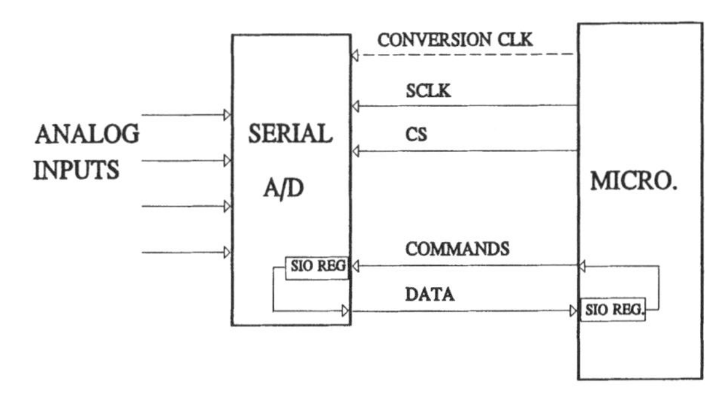

Figure 6.13 Typical serial ADC interface.

sion time. If not required, the two clock inputs can be tied together on the ADC. It is also possible to tie the A/D data in and data out together when the data in line are tristate after the required input bits are shifted.

When the chosen microcontroller does not provide a synchronous serial mode, the serial communication is much more difficult. An example occurs with the common 80C51 microcontroller. To communicate with this type of microcontroller, the above process must be carried out with bit manipulation instructions (also known as *bit banging*). This will significantly slow down the effective throughput compared to the more automated approach mentioned earlier and is due to the large number of instruction cycles while performing I/O manipulation.

Parallel interface

There are basically two ways to interface with a microprocessor—as memory-mapped or programmed I/O. The operation is nearly the same in both cases, however. In the memory-mapped approach, the external memory read and write control lines are also used to access I/O. This provides a more efficient transfer by allowing for all the memory accessing instructions to be used. With the programmed I/O technique, the microprocessor has additional read and write controls for I/O, as well as special instructions for accessing the I/O data.

Figure 6.14 is an example of a typical parallel interface. This shows the basic requirements for using the address lines for decoding the selection of the ADC. Once the proper ADC address is outputted, the read or write operation is activated. Depending on whether the read or write is active, the proper buffer/latch is enabled to transfer data to or from the ADC. When the conversion is complete, the ADC generates a "conversion complete" signal which will interrupt the processor.

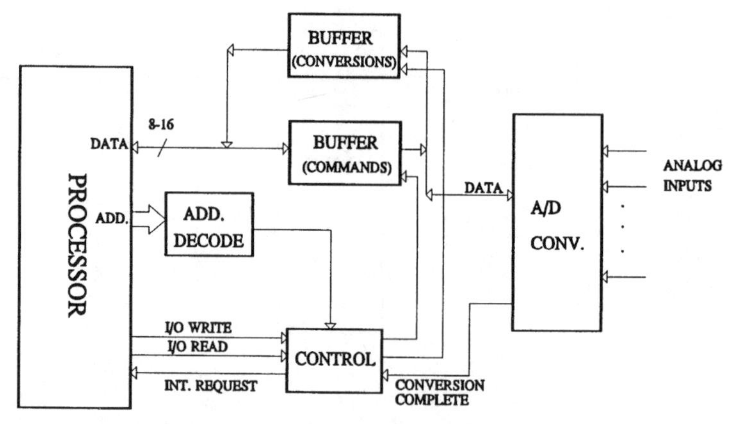

Figure 6.14 Typical parallel ADC interface.

This allows the processor to perform other tasks while the conversion is in progress, and it simplifies the timing.

Highly Integrated Converters

One of the major advantages of the successive-approximation converters being designed with CMOS technology is their ability to integrate digital circuitry. Besides the common features of a multiplexer, S/H, and reference, newer ADCs now can incorporate features that help off-load the processor.

Commonly integrated functions:

1. Self-calibration
2. Programmable modes (i.e., conversion rate, resolution, and bus width)
3. First-in first-out (FIFO) buffer memory
4. Direct memory access (DMA)

An example of self-calibration is the National Semiconductor ADC12451 12-bit converter. Similar devices actually contain either a small dedicated microcontroller or a state machine for automatically controlling the correction process. For the ADC12451, there are two correction DACs, as shown in Fig. 6.15. By incorporating internal DACs, the ADC12451 can be manufactured without trimming. This also provides higher performance in high-temperature applications with the ability to periodically perform recalibration. The correction DAC is used to correct for any offset voltages during the autozero calibration. During a full calibration cycle, nonlinearity errors (INL, DNL) are also corrected with this DAC. Correction values are then stored in internal RAM for automatically performing corrections while the main DAC is performing each measurement.

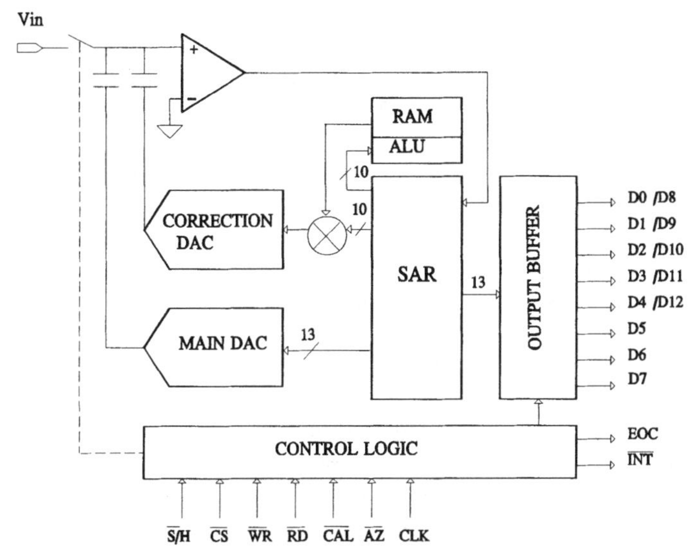

Figure 6.15 Example of a self-correction 12-bit ADC.

When an A/D converter such as the ADC12451 that provides 12 bits plus sign is used, it is possible to actually provide 13 bits of resolution for single-polarity inputs. To accomplish this, the input must be expanded to both positive and negative voltages. For example, if $V_{ref} = \pm 5$ V, and V_{in} full scale = + 5 V, then an amplifier/summer that subtracts 2.5 V and multiplies by 2 will work. In other words, 0 percent now equals − 5 V (V_{in} = 0 V), and 100 percent (full scale) equals + 5 V (V_{in} = + 5 V).

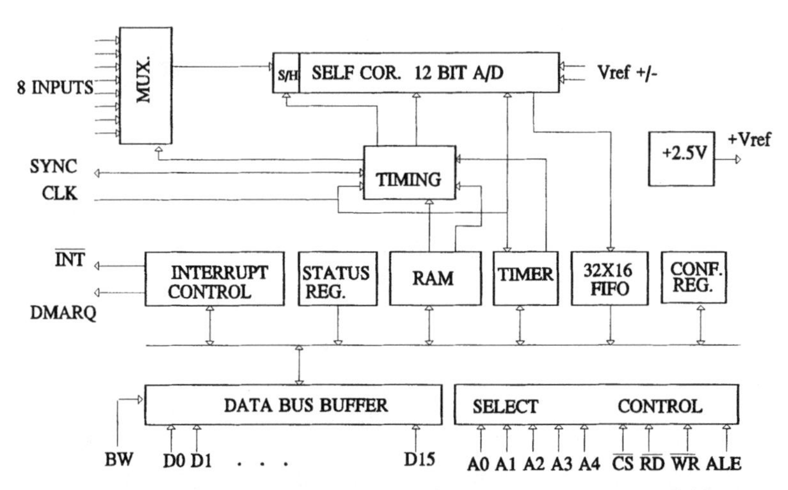

Figure 6.16 Example of a highly integrated data acquisition system (DAS).

Taking integration one step further, the National Semiconductor ADC12458 adds features 2 through 4 listed on the previous page. One of the most powerful features of this device is the 32×16 FIFO buffer memory shown in Fig. 6.16. The FIFO buffer stores consecutive conversions and can generate a DMA request (DMARQ) when the threshold is reached in the FIFO buffer. This allows the system processor and DMA controller to perform a DMA operation on several words, thus saving the large software overhead compared to an interrupt-based approach. Interrupt-based designs cause the processor to stop what it is doing, store the important parameters, and read each ADC result. This occurs for every conversion!

Another useful feature of the ADC12458 is its ability to program the acquisition time, resolution, and conversion time. This means that each channel can be customized for optimum performance. For instance, if some channels need to be accurate only to 8 bits, those channels can be programmed accordingly to speed up the total system conversion time. Other channels that have higher input impedance can have the acquisition time extended if necessary.

Chapter

7

Flash A/D Converters

When system designs call for the highest speed available, flash-type A/D converters (ADCs) are likely to be the right choice. Applications for flash converters include radar, high-speed test equipment, medical imaging, and digital communication. What differentiates the flash converter from other types is that input signals are processed in a parallel method. This chapter provides many insights as to how flash converters operate. In addition, many design tradeoffs are discussed not only in the architectures, but also in the process used for the converter (that is, CMOS/bipolar). We will take a look inside the main architectures so that potential error sources are identified and maximum performance is achieved.

Besides understanding how the various types function, it is necessary to consider the support circuits required. These primarily consist of buffer amplifiers, references, output latches, and memory interfacing. Due to the high speed associated with a flash ADC, a sample-and-hold (S/H) is usually not required. Understanding how the support circuits affect the operation of the ADC results will minimize performance problems.

Application circuits are presented to illustrate some commonly used techniques for performance enhancements. Some examples of how to improve the speed and power consumption and to lessen the speed requirement of the memory are presented. In addition, the final section covers some typical applications where flash ADCs are used.

Basic Architecture of Flash A/D Converters

Flash converters operate by simultaneously comparing the input signal with unique reference levels spaced 1 LSB apart. This requires several front-end comparators and a large digital encoding section. Simultaneously, each comparator generates an output to a priority encoder which then produces the digital representation of the input signal level. To accomplish this, it is necessary to utilize one comparator for each LSB increment. For instance, an 8-bit flash converter requires 255 comparators along with high-speed logic to encode the comparator outputs. This is somewhat of a brute-force approach to measuring the input signal. Consequently, the chip size and power consumption are sacrificed in favor of speed.

Figure 7.1 illustrates the basic flash ADC architecture, showing the front-end comparators with the resistor ladder network. Note that the input signal is simultaneously measured by each comparator which has a unique resistor-ladder-generated reference. This produces a series of 1s and 0s such that the outputs will be all 1s when the input signal is above the individual reference levels and all 0s when below the reference levels. The comparator output format is called a *thermometer* code because it resembles what you would see on a mercury thermometer.

Following the comparator outputs is the digital section consisting of several logic gates for encoding the thermocodes. The thermometer

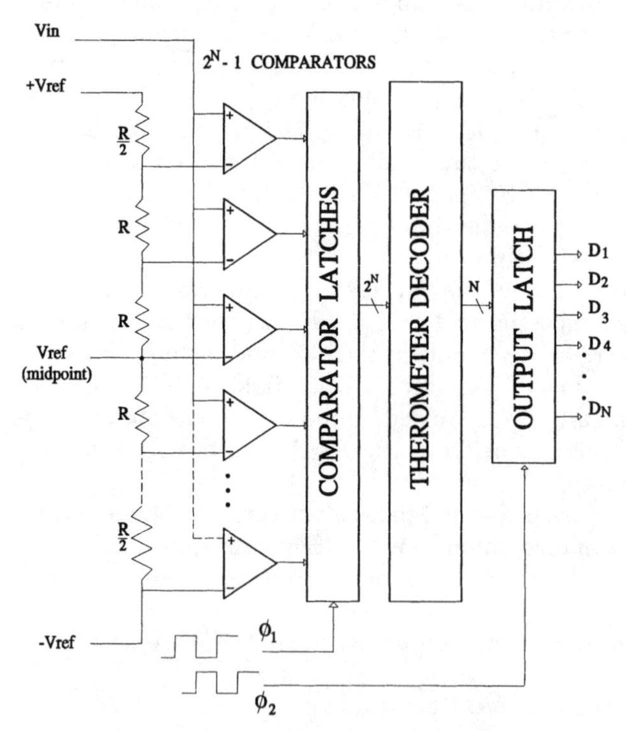

Figure 7.1 Basic architecture of a flash A/D converter.

decoder determines the point where the series of 1s and 0s form a boundary. The priority encoder uses this boundary threshold for conversion to a binary output. Outputs from the priority encoder (which are usually buffered in a latch) are then available to the system memory. It is important that the memory system be designed properly to prevent lost data since every new conversion will overwrite the previous result. For example, when the flash converter is operating at a rate of 100 MHz, new data must be stored every 10 ns!

Timing requirements

The type of flash converter architecture (and sampling rate) has an impact on the speed requirement of the memory. Flash converters without an internal latch require only one clock input. As an example, Fig. 7.2 shows the input being tracked during the low portion of the clock and the output latched during the high portion. To allow time for the outputs to settle, there will be some period in between the two phases where the output will be invalid. Essentially, the sampled input is delayed only about one-half a clock period (or equal to the sample duration). This requires that the memory write operation take place slightly less than one-half clock period (that is, 10 ns for a 50-MHz clock). In some situations, it may be necessary to extend the latch time to allow for slower memory.

Flash converters that utilize an internal latch greatly simplify the timing requirements. Operation with a two-phase clock provides control for the input sampling and the encoded output. Returning again to Fig. 7.1, note that the comparator outputs are controlled with a flip-flop (latch) using the phase 1 clock and that the output latch is controlled with the phase 2 clock. The input comparators continue tracking the input signal while the output latch holds the previous conversion result. This allows the comparator latches time to settle before the output latch is loaded and extends the valid data to one complete period. Figure 7.3 illustrates the timing relationship between the two phase-clocks. Note that there is a one-period pipeline delay as a result of the two phase-clocks. However, this is not really a problem since all signals are delayed equally. With the output valid

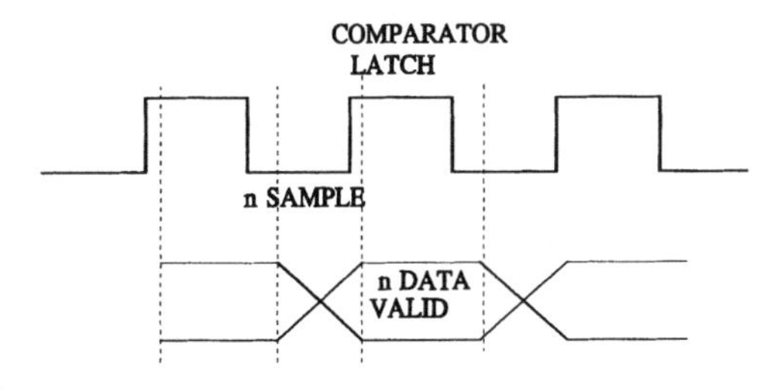

Figure 7.2 Single-phase clock timing.

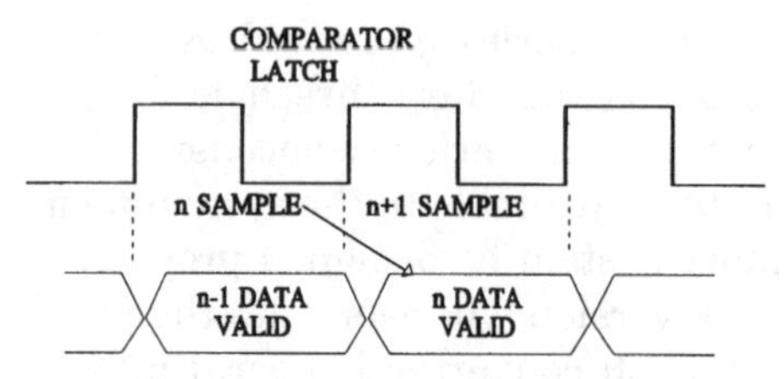

Figure 7.3 Flash converter two phase-clock pipeline delay.

for an entire clock period, the memory speed requirement is reduced by half.

CMOS Flash A/D Converters

There are several differences between the CMOS and bipolar flash converters. However, the largest difference is concerned with how the front-end comparators are created. CMOS technology allows for taking advantage of the ease of utilizing analog switches and capacitors. Figure 7.4 is an example of how a simple high-gain inverter with a set of analog switches and a capacitor is all that is necessary for creating a comparator. Basically, this is the same sampled-data comparator that was described in Chap. 6 for use in sampling and successive-approximation ADCs. The comparator works by first closing the feedback switch S_1 and connecting the reference input to the capacitor by also closing S_2 during phase 1. With the output fed back to the input of the inverter, the inverter input/output is forced to a state of equilibrium near one-half the supply voltage (both N- and P-channel output transistors are on). This initializes the capacitor to an autozero voltage of

$$\text{Autozero} = V_{c1} = \frac{V_{cc}}{2} - V_{\text{ref}}(x)$$

where $V_{\text{ref}}(x)$ is the ladder network reference.

During phase 2, the switches are all reversed to open the inverter feedback and connect the input signal to the comparator. Since the

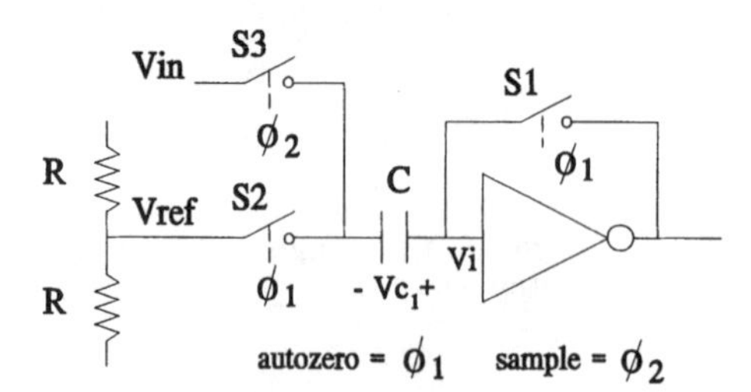

Figure 7.4 CMOS flash A/D comparator.

capacitor voltage will essentially stay the same, any difference between the input signal and the reference level will cause the comparator input to shift. The comparator input voltage change ΔV_i is described by

Phase 2 (input sample)

$$\Delta V_i = V_{ref}(x) - V_{in}$$

If the input signal V_{in} is less than $V_{ref}(x)$, the comparator input will be forced below the level necessary for equilibrium. In other words, the comparator output is forced to go high. The opposite is true when the input signal is higher than $V_{ref}(x)$.

There are several advantages to creating a comparator as described above. First, the input offset voltage normally associated with a two-terminal comparator is not an issue since the inverter has only one input. The actual comparator input voltage V_i during the autozero phase is not important either. This is because operation of the comparator is dependent only on the capacitor to maintain the difference between V_i and the reference level. In addition to eliminating initial dc offset voltages, the self-bias point of the inverter-type comparator is not affected by temperature or time.

Contrary to what some may believe, CMOS flash converters can equal the speed of all but the ultrahigh-speed bipolar designs with emitter-coupled logic (ECL). However, there is a big difference in power consumption between CMOS and bipolar converters. Since CMOS designs have both N and P channels, the power consumption will be dependent on the frequency of operation or when both the high and low sides are simultaneously on. As will be explained later, power is mainly consumed during the autozero mode when the inverter (comparator) is driving both the N- and P-channel output transistors.

Error sources of CMOS flash A/D converters

Many of the potential error sources are common to both CMOS and bipolar flash A/D converters. However, some error sources are unique because of the differences in comparator designs. For instance, the CMOS sampled-data comparator design has similar switched-current problems, as mentioned in Chap. 6. This current will vary along the string of comparators due to the difference between the input signal and various resistor ladder levels. When V_{in} is near 0 V, the net switched current will have a positive amplitude. When V_{in} is closer to V_{ref}, a net negative amplitude is produced. With V_{in} approximately equal to $V_{ref}/2$, there is a cancellation effect because the net capacitor

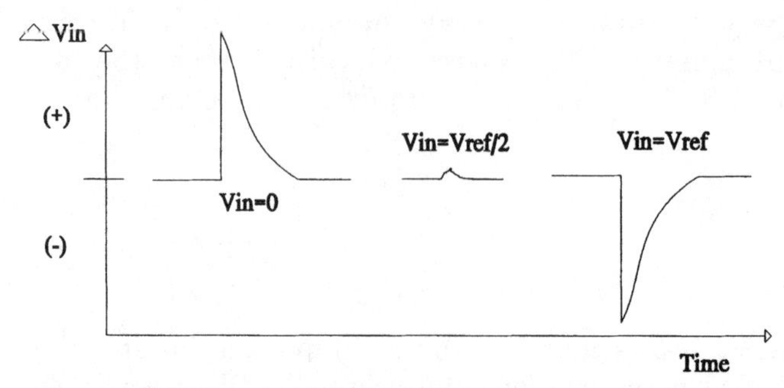

Figure 7.5 CMOS flash ADC switched input currents.

charges on comparators above V_{ref} equal the net charge on comparators below V_{ref}. Figure 7.5 shows the switched currents.

The potential problem with the switched input currents is that the input voltage must settle within the conversion time of the ADC. The settling time is worst when the input signal is at the limits of the input range. To determine the settling time, you can lump together all the capacitance values of the A/D input stages C_t and the buffer amplifier output resistance with the analog switch resistances R_t. After the autozeroing operation, the net voltage on the capacitor array is approximately $V_{ref}/2$. As an example, assume that a 10-bit ADC is required to settle within ½ LSB, and let $R_t = 100\ \Omega$ and $C_t = 30$ pF. The worst-case settling time is computed as follows:

$$V_{in}(f_s) = \frac{V_{ref}}{2\,e^{-T_s/(R_tC_t)}} = \frac{V_{ref}}{1024}$$

Solving for T_s gives

$$\begin{aligned} T_s &= R_tC_t \ln 1024 = 100 \times 30 \text{ pF} \times 6.93 \\ &= 20.8 \text{ ns (approximately} = 50 \text{ MHz)} \end{aligned}$$

Besides the settling time, there is another problem associated with the switched capacitor network. Unavoidable parasitic capacitance from the switch gates to the source/drain causes charge feedthrough to the sampling capacitor. Although it is not really an offset voltage, it has the same effect.

Bipolar Flash A/D Converters

As mentioned above, the largest difference between the CMOS and bipolar flash ADCs is in the front-end comparator stage. Due to the radical difference in the input stages of CMOS and bipolar flash converters, the frequency-response limitation is also quite different.

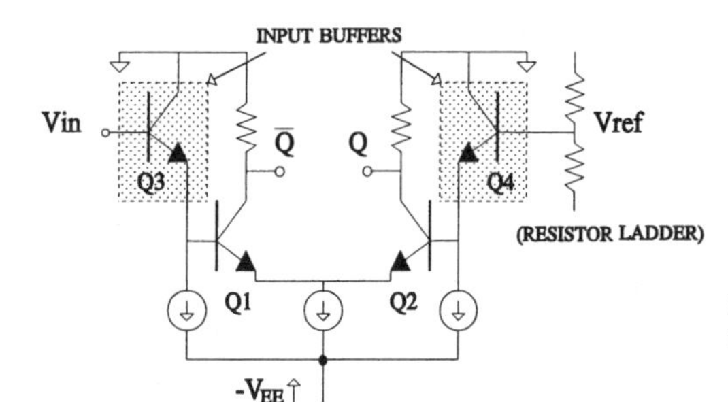

Figure 7.6 Bipolar flash A/D differential comparator.

Figure 7.6 shows a typical bipolar comparator. Transistors Q_1 and Q_2 form the differential pair, and Q_3 and Q_4 are emitter-follower buffers. Buffers are required as part of the bipolar comparator to prevent the input and reference signal from excessive comparator loading. The frequency response of the input buffer stage is largely responsible for the dynamic performance limits of the ADC.

Depending on the speed requirements of the bipolar flash converter, the logic section can be either standard TTL/CMOS level or ECL. Also note in Fig. 7.6 that this design operates with a negative supply voltage. The highest speed possible is obtained by using ECL for the encoder stage which requires a negative supply voltage. With a negative supply voltage for the encoder, the bipolar comparators also must operate from a negative supply. Designs that provide only TTL/CMOS interfaces typically have the advantage of operating from a single + 5-V supply. ECL circuit operation is faster because it keeps the logic transistors from operating in the saturated state (restricted to either cutoff or active). This eliminates the charge storage delays that occur when a transistor is driven in the saturated mode.

Error sources of bipolar flash A/D converters

Due to the nature of the bipolar comparator input stage, there are other potential error sources. Referring again to Fig. 7.6, note that this is a two-terminal comparator where any mismatch in transistors Q_1 through Q_4 will cause a net V_{os}. This dc error will also be temperature-dependent. Furthermore, mismatches will affect the uniformity of the reference taps and cause nonlinearities (decreased *S/N*).

Another problem with the differential bipolar comparator is the nonlinear effect of the input capacitance. Unlike the CMOS sampled-data comparator, the bipolar comparators exhibit a junction capacitance from the parasitic p-n junctions. The resulting input capacitance will follow the nonlinear curve of capacitance versus reverse-biased voltage. In addition, this capacitance is affected by temperature. Both the reverse-biased voltage and the temperature have an inverse square root effect on the input capacitance. Therefore, the performance is more difficult to predict and correct for.

Encoder Requirements

Up to now, we have been discussing only the front-end comparator stage. The comparator outputs normally produce the expected thermocode where there is a consistent string of 1s and/or 0s. However, when the input signal approaches one of the converter's references, the very small overdrive may cause the comparator to slew too slowly. When high-speed flash converters sample the input signal that is near a reference level, the comparator output could potentially be an indeterminate state. If this condition exists long enough and/or the sampling frequency is too high for reliable operation of the latch, a condition known as *metastability* exists. The worst-case scenario occurs when the MSB is affected since this will cause gross errors, as shown below. Depending on the location of the metastable state, the error magnitudes will be different.

Metastable example:

$$\text{True code} = 1000000000 \qquad \text{10-bit A/D code}$$

$$\text{Actual code} = 0000000000$$

The above represents a gross error that is often referred to as a *sparkle code*; this is what would be seen if the input were a video signal. One way to correct for the sparkle codes is to use additional logic for the encoder. For example, a Gray code will guard against three adjacent comparators having inconsistent outputs (that is, 101) and will produce a maximum of 1-LSB error. However, before the encoded logic signal can be used, first it must be converted to binary. As usual, nothing comes for free, and this extra logic will tend to limit the speed of the converter. For this reason, designers of flash converters often use a "pseudo-Gray-code" technique to reduce the digital delays.

Sampling Jitter Errors

The *aperture jitter* T_{aj} is the sample-to-sample variation that can result from timing differences within the flash converter or externally in the sampling clock. This variation will degrade the signal-to-noise (S/N) ratio and increase the harmonic distortion of the ADC. Internally generated jitter can result from mismatched timing delays of either analog or digital circuitry. To get an idea of the performance limitation of the ADC due to sample jitter, let's take a look at an example. Let

$$V_{in} = 2V_{p\text{-}p} \text{ sine wave} \qquad T_{aj} = 10 \text{ ps}$$

$$\text{Desired resolution} = 10 \text{ bits}$$

Determine the maximum frequency for 1-LSB error:

$$\frac{dV_{in}}{dt} = \frac{d(V_p \sin wt)}{dt}$$

$$= 2\,(3.14)\,(f) \cos wt$$

The maximum rate of change equals the zero crossing where cos $wt = 1$.

$$\frac{dV_{in}}{dt_{max}} = \frac{2}{1024\, t_{max}} = 6.28\, f_{max}$$

$$f_{max} = \frac{2}{1024 \times 6.28 \times 10 \text{ ps}} = 31.1 \text{ MHz}$$

If the input signal exceeds the frequency calculated above, there will be missing codes in the result.

Subranging Flash A/D Converters

When high speed is still desired but it is also important to minimize cost and power consumption, a subranging (or multistep) type of flash converter is a potential solution. This adaptation of the conventional flash ADC requires far fewer comparators, less encoder logic, and less power. The term *subranging* is used because the total conversion requires two or more estimates that reduce subsequent conversion ranges. In a sense, the successive-approximation and variable-range pulse-width modulators are also subranging converters since both use estimates of the input for succeeding measurements.

Half flash converter

As an example of the basic idea of subranging, the half flash converter architecture is examined. The converter in Fig. 7.7 is referred to as a *half flash* converter because there are two steps in the conversion process. Each step produces half the number of bits of final resolution. Note that the first conversion is a flash conversion on the input and presents the most significant portion of the result to the output. Next, a DAC converts the estimate so it can be subtracted from the original analog input for the final conversion step. The second flash converter now completes the process by performing a flash conversion on only a small portion of the full range. For example, in an 8-bit con-

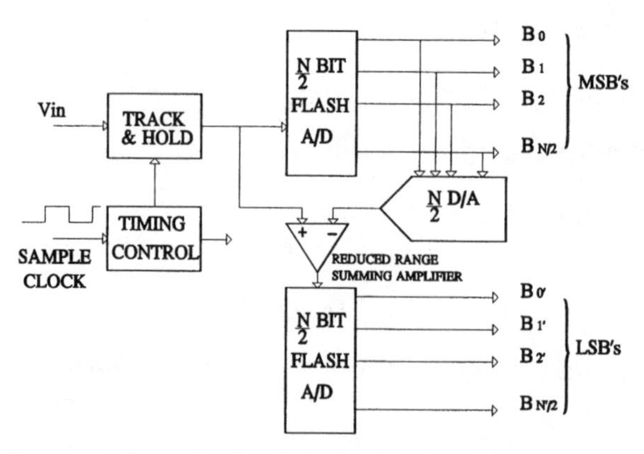

Figure 7.7 Example of an N-bit half flash architecture.

verter, the second conversion will perform a conversion on only one-sixteenth of full scale for generating 4 LSBs of the result.

The number of comparators required for a full flash converter and a half flash converter are determined as follows:

$$\text{Full flash comparators} = 2^N - 1$$

$$= 255 \text{ for } N = 8$$

$$\text{Half flash comparators} = 2(2^{N/2})$$

$$= 2 \times 16 = 32 \text{ for } N = 8$$

In a practical design, the extra hardware steps (i.e., DAC, summer) would likely produce additional unwanted errors. To compensate for this, a digital correction technique is often used. Essentially, this technique uses more resolution in the final flash conversion (lower byte) than is required in the final result. The unused LSB is then fed back to the MSBs to zero out induced errors. Therefore, compensation is made for errors in the first flash estimator (A/D), D/A, summer amplifier offset, and the second flash converter voltage offset.

Multistep flash converter

Expanding the half flash concept further will reduce the number of comparators and required encoding logic even more. Benefits of the half-step or multistep techniques become more significant when the resolution is 10 bits or greater. For example, a 10-bit full flash converter requires 1023 comparators, the half flash converter still requires 64 comparators, and a multistep flash converter requires about 16 comparators. Both cost and power reductions are realized by applying the multistep technique.

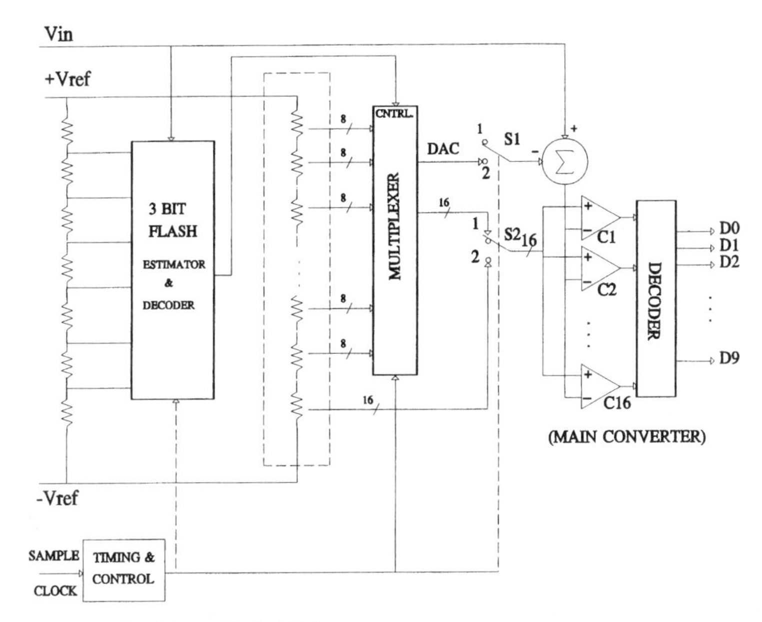

Figure 7.8 Multistep flash ADC.

As an example of how the multistep technique can be implemented, we examine the National Semiconductor ADC1061 10-bit converter. In Fig. 7.8, switch positions 1 and 2 indicate the two steps required. The front-end comparator bank performs a 3-bit estimate similar to that described for the half flash converter. However, this is where the similarity ends. During the first conversion step, an additional step is performed. Based on the input estimate, the second comparator bank (main converter) will perform a flash conversion with the 16 required resistors overlapping the first-estimate range. This ensures that errors in the front end will not affect the overall accuracy. The results of the first conversion represent the MSBs. To complete the process, S_1 and S_2 are connected to position 2. This switches in the resistor tap voltage just below the estimated input from the input stage. Simultaneously, the lower 16 resistor taps are switched to the output main converter. Therefore, the next flash conversion completes the LSBs of the result. With this approach, the number of comparators required is

$$\text{Multistep comparators} = 2^{(N/2\,-\,1)} + \text{estimator}$$

$$= 8 \qquad \text{for } N = 8$$

$$= 16 \qquad \text{for } N = 10$$

Support Circuits

Support devices for flash ADCs are required to buffer and amplify the input signal, drive the resistor reference ladder (reference), and interface the converter to the memory system. Flash converters require special care due to the very high-speed operation, input capacitance, and low-impedance resistor ladder network. When you are choosing support devices, make sure that the errors introduced are less in magnitude than the ADC errors. It is always a good idea to refer to the manufacturer's data sheets for recommendations on proven support circuits to save valuable design and test time.

Buffer and amplifiers

Figure 7.9 illustrates a typical buffer/amplifier circuit. In addition to the components shown, power supply decoupling capacitors (large electrolytic and small ceramic) located close to the amplifier supply pins are a must. The buffer and amplifier not only provides the necessary drive current, but also inverts and scales the input signal to match the typical converter input range of 0 to -2 V. Offset correction is an option, and this can be accomplished with either the potentiometer circuit shown or digitally with the system processor.

Resistor R_s is used to isolate the converter input capacitance from the amplifier to prevent pulse response overshoot. The resistor also helps limit the current through the diode which is used for preventing positive inputs from damaging a unipolar flash converter. Keep in mind that this series resistor and the internal capacitance form a lowpass filter. For this reason, the 3-dB frequency of the *RC* network

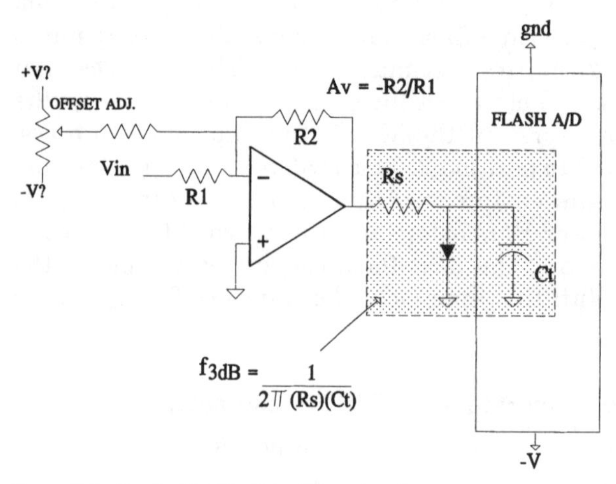

Figure 7.9 Typical buffer/amplifier circuit.

must be set sufficiently higher than the maximum input signal frequency (or the bandwidth of the ADC). For example, when a 20-MHz, 8-bit flash converter with 20-pF input capacitance is used, the maximum acceptable total series resistance can be calculated as follows:

$$f_{3\,\mathrm{dB}} = 5 \times 20\ \mathrm{MHz} = 100\ \mathrm{MHz}$$

$$R_s = \frac{1}{2\,(3.14)\,(910\ \mathrm{MHz})\,(20\ \mathrm{pF})} = 79\ \Omega$$

Note: Keep in mind that the series resistance R_s includes the amplifier output impedance which will increase with frequency.

The amplifier selection should be made by closely evaluating the application to determine the important specifications. In Chap. 3, requirements for buffer and amplifiers were discussed to make this decision easier. Specifications to consider in choosing buffer and amplifier include the bandwidth, slew rate, drive current, distortion, and dc performance. Perhaps the most important specification in the selection of an amplifier for a flash converter is the bandwidth. To go along with bandwidth, the gain flatness, peaking, and roll-off should also be examined. Depending on the desired resolution, the 3-dB bandwidth of the amplifier can be used in the following equation to determine if there is sufficient gain versus frequency [$A(f)$].

$$A(f) = \frac{1}{\sqrt{1 + (f_{\mathrm{in}} / f_{3\,\mathrm{dB}})^2}}$$

Reference circuit

Depending on the flash converter, reference circuits must provide either single- or dual-polarity voltages. Regardless of the type of converter, a low-impedance driver is essential to prevent the internal ladder resistance (which is typically 75 to 300 Ω) from affecting the reference voltage. The reference circuit in Fig. 7.10 shows a dual reference. Low-impedance drive is provided with small-signal discrete transistors at the amplifier outputs. Offset voltages are possible due to high current through the bonding wires to the converter's resistance ladder, and the offset voltages can be adjusted out with trim potentiometers. However, this may not be adequate since the internal voltage drops will change according to the temperature. To compensate for this, some converters utilize a drive and sense taps at the endpoints of the network. If additional taps are provided (i.e., half or quarter taps) for actively driving the resistor ladder network, then integral linearity errors from resistor mismatches and comparator base currents will be reduced.

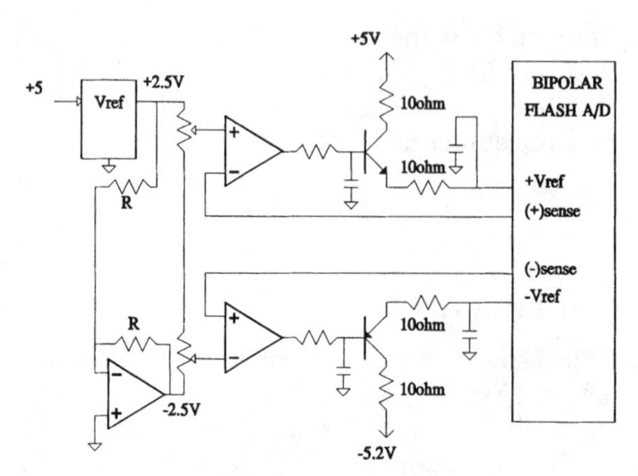

Figure 7.10 Dual-polarity reference circuit.

Maximizing Flash A/D Converter Performance

Although flash ADCs generally offer very high performance, there are several techniques that can be applied to improve the total system design. One reason for this is due to the parallel architecture of the converter. The improvements that can be made include higher resolution, higher speed, lower power, and reduced memory speed.

Stacking (increased resolution)

Providing the chosen flash ADC has an overflow output, two of the converters can be stacked as shown in Fig. 7.11 to effectively double the resolution. This works by connecting the negative reference tap of the upper converter to the positive reference tap of the lower, and using the overflow bit from the lower converter to disable itself in the event of an overflow. When the overflow occurs in the lower converter, the MSB is set and the remaining bits are provided by the upper converter.

Ping-Pong (double speed)

In applications where it is desirable to double the speed of the converter, the Ping-Pong technique can be a solution. With this approach, two converters are connected in parallel as shown in Fig. 7.12. Note that the clocks and output controls are 180° out of phase. This configuration alternately provides a result from each converter on every phase of the clock. However, when this technique is used, the errors from the two converters will add. Furthermore, any mis-

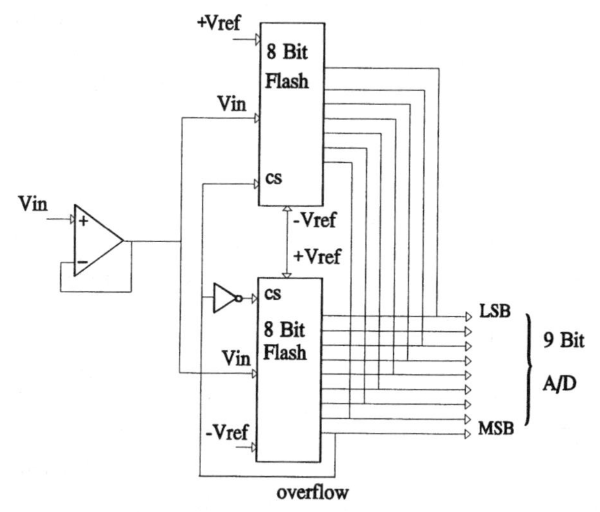

Figure 7.11 Stacking two flash ADCs for double the resolution.

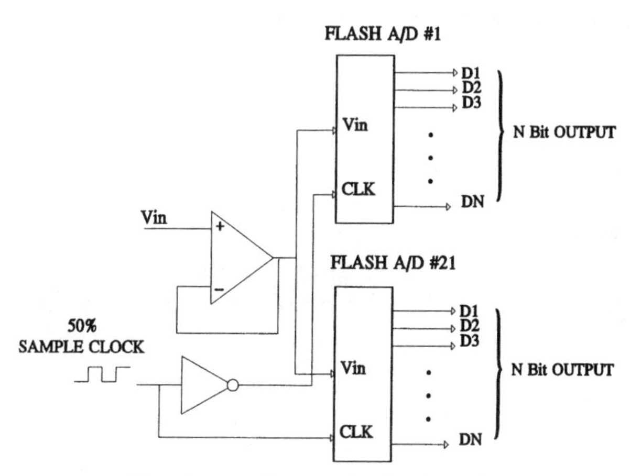

Figure 7.12 Ping-Pong technique for doubling the conversion rate.

matches in linearity or resistor ladder taps can produce jumps in consecutive codes.

Power reduction of CMOS flash A/D converter

Due to the nature of CMOS design, power consumption occurs only during clock transitions. Providing the application does not require either a maximum conversion rate or several consecutive measurements, there are techniques to minimize the power consumption. In CMOS flash converters, maximum power consumption occurs during the autozero mode, where the sampling comparator is forced into the linear mode. By reducing the duty cycle of the autozero time versus the conversion clock pulse width, the power will be minimized. This can be accomplished either by using two different duty cycle clocks (autozero and convert) or by performing a burst-mode conversion. In the burst mode, there are limited consecutive conversions completed followed by a relatively longer inactive period. Both techniques are illustrated in Fig. 7.13.

To reduce the power in CMOS converters, another possibility is to lower the power supply voltage. However, this will affect the conversion rate, so the manufacturer's data sheets should be consulted. Bipolar flash converters generally are more susceptible to problems if the supply voltage is reduced. This is due to the required bias voltages for the bipolar transistors. In systems where there are long periods in between measurements, maximum power savings are made possible by placing the CMOS converter in the standby mode. In standby, the CMOS converter power consumption will be in the microwatts range.

Bus interfacing

High-speed flash converters can pose problems for the system designer. First, the high-speed digital signals present on the bus can work their way into the analog and digital sections of the converter. Using an external buffer/latch will go a long way toward avoiding this problem. Second, how should the system designer handle the very high data rates from the output latches? In these applications, some form of data buffering will likely have to take place so that slower (more

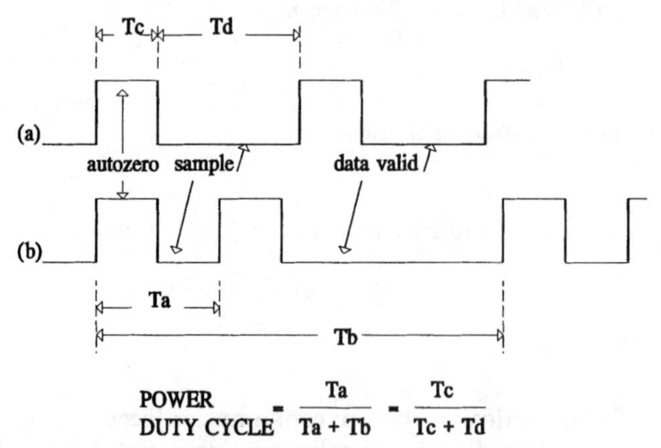

Figure 7.13 CMOS flash A/D power reduction. (*a*) Limiting duty cycle; (*b*) burst mode.

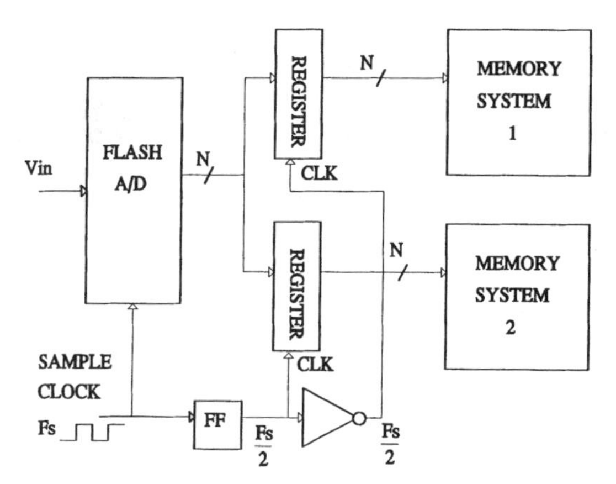

Figure 7.14 Data rate reduction with 2:1 demultiplexing.

economical) memory can be used. One way to accomplish this is to use a 2:1 demultiplexer, as shown in Fig. 7.14. Note that the output latches are clocked by half the sampling clock rate and that each is 180° out of phase. Some of the high-speed flash ADCs are available with similar demultiplexer logic built in, to simplify the interface.

Flash A/D Converter Applications

There are several applications that require the speed of a flash ADC. Among these are communication, instrumentation, and imaging HDTV (high-definition TV), and medical diagnostic equipment. With the aid of high-speed digital signal processing (DSP), the flash ADC outputs can be processed to bring almost limitless versatility to an application. The basic building blocks of a typical system with a flash A/D converter include

- Filter for analog input
- Flash ADC
- DSP and memory system
- Processor and interface system
- Digital-to-analog converter (DAC)
- Filter for analog output

Communication example

High-speed fiber-optic communication requires the analog signals to be converted to digital with a flash ADC. Following this, the flash

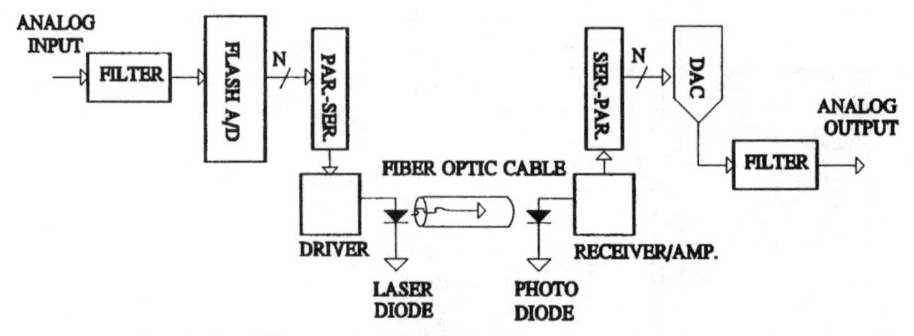

Figure 7.15 Example of fiber-optic communication.

ADC outputs must be converted from parallel to a serial output format for the driver to modulate a laser diode (Fig. 7.15). This allows a single fiber-optic cable to transmit several high-speed signals if necessary. On the other end of the cable, the photodiode detects the digital light pulses, which are then converted to logic levels in a parallel format. The DAC output is finally filtered by an analog filter. Depending on the application, an option may be to first perform digital filtering before outputting to the DAC.

Digital oscilloscope

A block diagram of a digital storage oscilloscope is shown in Fig. 7.16. Two input channels are first conditioned by a programmable amplifier. The trigger circuit controls when the sampling starts, which is controlled by the sample-and-holds. Buffer amplifiers are required to isolate the low-impedance flash A/D input stage from loading the sample-and-holds. Once digitized by the flash ADC, the input signals are stored in memory where a DSP can perform several operations (i.e., filtering, averaging, peak-to-peak time versus voltage, root-

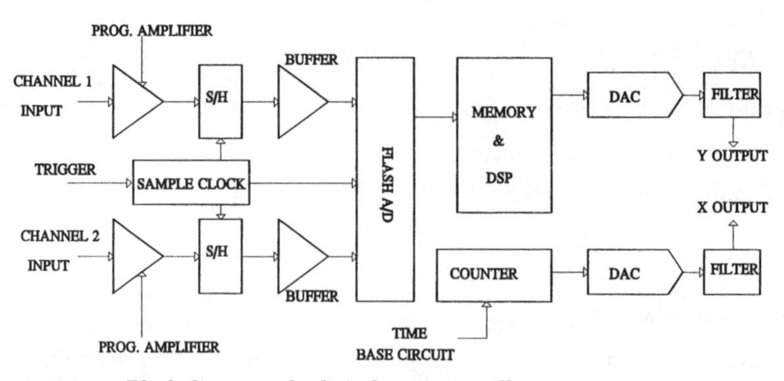

Figure 7.16 Block diagram of a digital storage oscilloscope.

mean-square, etc.). Two DACs then provide both the vertical (signal measurement) and horizontal (time measurement) analog output signals. Finally, the analog outputs are filtered before they go to the display driver.

Medical imaging

One form of medical imaging is digital x ray (radiography), illustrated in Fig. 7.17. With this technique, the images no longer need to be produced or stored on film. Instead, the images are stored within memory for display on a video monitor. In this system the x ray passes through a person onto a photomultiplier fluoroscope sensor which converts the x rays to light. The various light intensities are then converted to an analog signal which is filtered before being digitized by the flash ADC. The DSP then enhances the image before driving the output DAC for the video image. The advantages to this process are that (1) the time exposure to the x ray for a quality image is significantly shortened and (2) stored images can be manipulated to enhance specific areas.

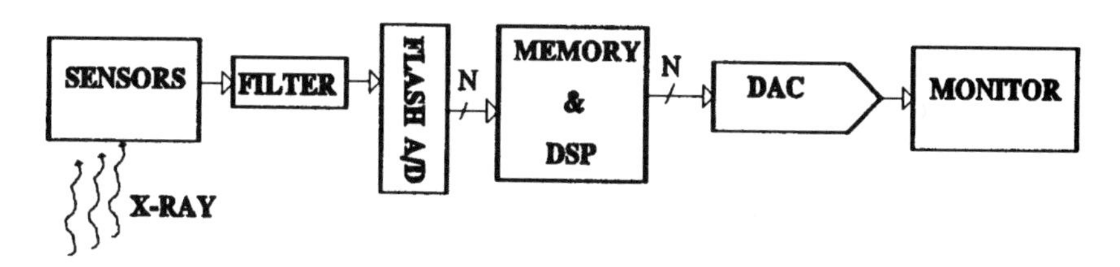

Figure 7.17 Block diagram of a digital radiography system.

Chapter

8

Delta-Sigma (Oversampling) A/D Converters

Oversampling converters are more commonly called *delta-sigma converters.* Other names include either *sigma-delta* or *charge-balancing* A/D converters (ADCs). Delta-sigma converters differ from other ADC approaches by sampling the input signals at a much higher rate than the maximum input frequency. Traditional, nonoversampling converters such as successive-approximation ADCs perform a complete conversion with only one sample of the input signal. In addition, the maximum sampling rate of successive-approximation converters is typically around 4 times the maximum input frequency. Another unique characteristic of the delta-sigma converter method is that a closed-loop modulator is used. The modulator not only continuously integrates the error between a crude ADC and the input signal, but also attenuates noise. This combination of oversampling and closed-loop modulation creates a very powerful technique.

Although the delta-sigma converter concept has been around for several years, it was not until the last few years with advances in VLSI technology and intensive research that practical solutions materialized. The reason to use this unique technique is to facilitate easy-to-manufacture digital circuitry along with low-precision analog circuitry. This allows for a highly integrated analog-to-digital converter to be created primarily with digital techniques. Shifting the accuracy and performance burden to the digital domain creates several advantages. For instance, filtering can be performed accurately with immunity to time and temperature drifts, compared to a more analog approach. In addition, there will be no phase error associated with a digital filter, and it is relatively easy to produce high-order attenuation of almost any order desired. The characteristics can also be modified "on the fly" under software control.

Concepts used for creating the delta-sigma ADC can also be applied to creating a digital-to-analog converter (DAC). Digital audio systems, e.g., require both an ADC and a DAC with digital filtering in between for performance enhancement. The delta-sigma method can be extended to create a high-performance DAC by applying a technique referred to as *interpolating*. This basically increases the output frequency of a low-resolution analog output (DAC). The net effect significantly improves the resolution and makes the DAC output easier to filter.

Filter Requirements Due to Aliasing

To get an appreciation for the delta-sigma method, let's first look at the conventional sampling ADC, and its associated difficulties. Figure 8.1 shows a conventional sampling ADC that is sampling at the Nyquist rate. The sampling theorem states that in order to accurately reproduce an input signal, that signal must be sampled at no less than 2 times its highest in-band frequency F_b. This is referred to as the *Nyquist rate.*

$$\text{Nyquist sampling rate } F_s = 2F_{b\,\max}$$

Aliasing is a phenomenon which can corrupt in-band signals when the sampling rate is not sufficiently high compared to F_b. In Fig. 8.2 note that the bandwidth of these signals is F_b and is centered at integer multiples of the sampling frequency F_s. Low-pass filtering must

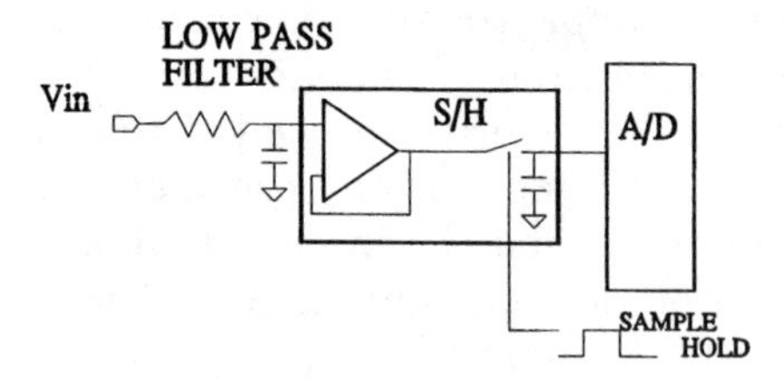

Figure 8.1 Nonoversampling ADC.

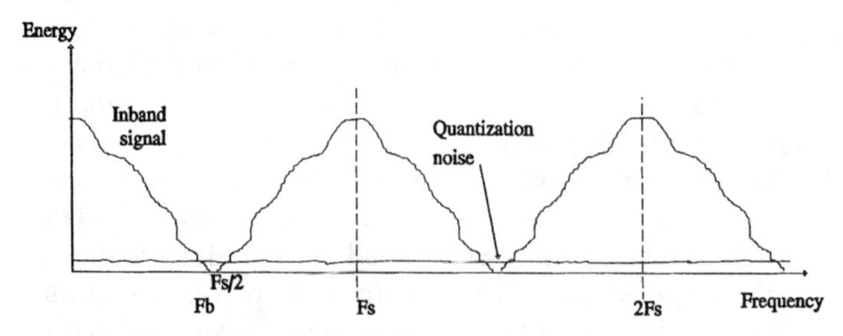

Figure 8.2 Frequency spectrum input, noise, and aliasing.

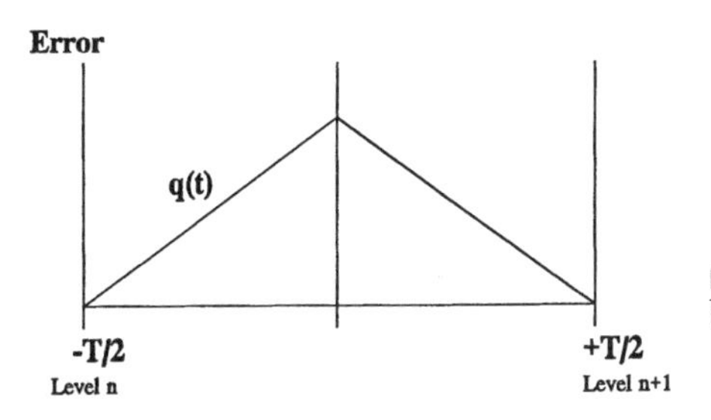

Figure 8.3 Quantization noise between digital codes.

then precede the sampling process to ensure that the attenuation of the input frequencies (or noise) above $F_s/2$ does not cause an overlap with the harmonics. The filter must also provide sharp attenuation (high order) so the in-band signals are not undesirably restricted. This requirement can be difficult and expensive to implement, especially for high-resolution data conversions. Another drawback to the nonoversampling method is that the front-end low-pass filter can introduce significant errors and can take up considerable board area. This will likely be the situation when more than a simple *RC* low-pass filter is required for high-resolution ac signal measurements.

Quantization Noise

Quantization noise is present in all analog-to-digital converters and is due to the finite resolution in measuring a "busy" input signal. This noise reduces the *effective number of bits* (ENOB) and the dynamic range (usable resolution) of the ADC. It will become apparent later in this chapter that reducing the quantization noise is vital to the delta-sigma process. Typically, an analog-to-digital converter is listed as having ± ½-LSB resolution of the quantizer for a nonoversampling converter. The minimum error is not ± 1 LSB because of the centering of the ADC thresholds between the true signal thresholds.

Understanding how quantization noise is created and minimized is central to the use of the delta-sigma method. When an input signal is at a threshold of the ADC, any changes in amplitude will cause what appears to be noise in the digital codes. The noise will have equal probability in frequency (uncorrelated) across the spectrum and will resemble white noise, as shown in Fig. 8.2. In Fig. 8.3, you can see that the effective error and noise will increase from zero at the thresholds ($T/2$) in a linear fashion as the input signal approaches the midpoint between digital codes.

Mathematical derivation of quantization noise:

$$E^2_{\text{rms}} = \int_{T/2}^{-T/2} \frac{(qt)^2}{T}$$

$$E^2_{\text{rms}} = \frac{q^2t^3}{3T^2}\Bigg|_{-T/2}^{+T/2}$$

$$\text{Noise level as function of frequency} = \frac{q^2}{12F_s^2}$$

where F_s = sampling frequency. Taking the square root of both sides yields an error e of

$$e = \frac{q}{\sqrt{12F_s}}$$

From the above equation showing the noise level as a function of frequency, it can be shown that oversampling will reduce this noise level. For every doubling of the sampling frequency F_s, the resolution is increased by 0.5 LSB. By sampling at a much higher frequency than the Nyquist rate, noise is effectively spread out over a wider frequency range. The following equation describes the effect of oversampling:

$$\text{In-band noise power} = \int_{-f_n}^{+f_n} N(f)\, df$$

$$= \frac{2F_n q^2}{12F_s}$$

Similarities Between PWM A/D and Delta-Sigma Converters

The improved PWM A/D technique described in Chap. 4 is an example of one of the many similarities to the delta-sigma concept. This will help develop an intuitive understanding of how the delta-sigma method works. The PWM A/D configuration is revisited with an added flip-flop for synchronizing with a sampling clock. In actuality, the flip-flop is part of the microcontroller. In Fig. 8.4, note that the comparator compares the sum of previous outputs stored on the capacitor at one input and the input signal at the other input. Each output from the comparator is then sampled at regular intervals with the sampling clock F_s on the flip-flop. Multiple high-frequency sam-

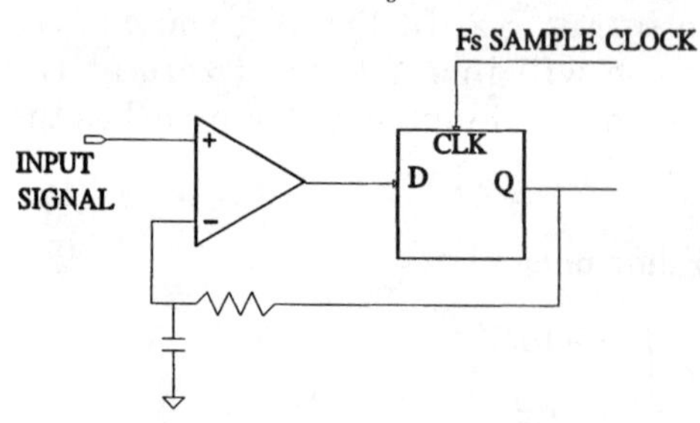

Figure 8.4 Example of a PWM A/D converter with sampling clock F_s.

ples are taken from the comparator output for determining the drive pulses required by the *RC* integrator. This closed-loop action will force the accumulated "guesses" of the low-resolution ADC to equalize the unknown input signal. Like the delta-sigma ADC, the PWM A/D converter will force the voltage on the *RC* integrator to equal the input voltage over a period of samples. The only way the inputs of the comparators can be made equal is for the average time-weighted voltages of the high-frequency pulses to equal the unknown input signal.

The main function of the microcontroller is to provide closed-loop control by reading the comparator and then driving the *RC* circuit with a single-bit DAC output. In addition, the microcontroller performs crude digital filtering. This filtering can be considered a basic *finite impulse response* (FIR) filter that merely computes the running average of several measurements. The equation and simple example below illustrate the basic operation:

$$V_{in} = \frac{1}{N \Sigma (X - i)}$$

where $i = 0$ to $N - 1$
$X =$ quantized reference (1 or 0)
$N =$ number of samples

Clock	DAC level (X)	
0	1	
1	1	
2	0	
3	1	$N = 8$
4	1	
5	0	
6	1	
7	+ 0	
	Sum = 5	

Delta-Sigma A/D Converters

Delta-sigma ADCs operate in much the same way as described above (improved PWM ADC). However, this class of converter provides increased performance by using some additional stages. Previously we described that the effects of quantization noise can be reduced by oversampling the input at a high rate. In order to get significant benefits from oversampling alone, it is usually necessary to sample at a very high rate. This can be impractical for many applications. By adding a step to the process, the benefits of oversampling can be more fully realized. This is where the method known as *delta-sigma modulation* comes in. Delta-sigma converters quantize the difference between the current signal and the sum of previous differences by using an integrator (usually two or more) within the closed loop. Although this is similar to the RC integrator, the delta-sigma modulators with multiple active integrators increase performance significantly, the most notable of which is the conversion time.

Figure 8.5 shows the basic block diagram required for performing the delta-sigma process. Note that the RC integrator has been replaced by an active integrator. The integration of the errors between the input and DAC output forms a closed loop with the output fed back to the input differential amplifier. This servo action will force the average quantized output to equal the input over time. In other words, a running average of the low-resolution guesses is taken by using the difference between the input and guesses of the input.

To illustrate the power of the delta-sigma converter, it is helpful to refer to Fig. 8.6, which shows an addition of noise to the model. This model is for explanatory purposes only and shows how quantization noise affects the system and how it can be controlled.

The signal transfer function can be illustrated by

$$Y(s) = [X(s) - Y(s)]\frac{1}{s}$$

with $N(s) = 0$:

$$\frac{Y(s)}{X(s)} = \frac{1}{1 + s}$$

with $X(s) = 0$:

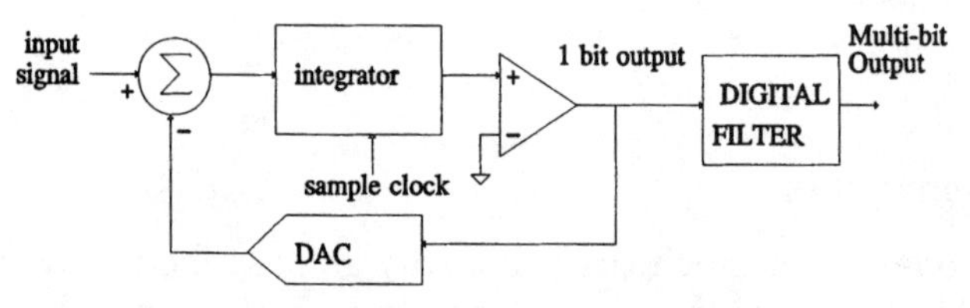

Figure 8.5 Basic block diagram of delta-sigma ADC.

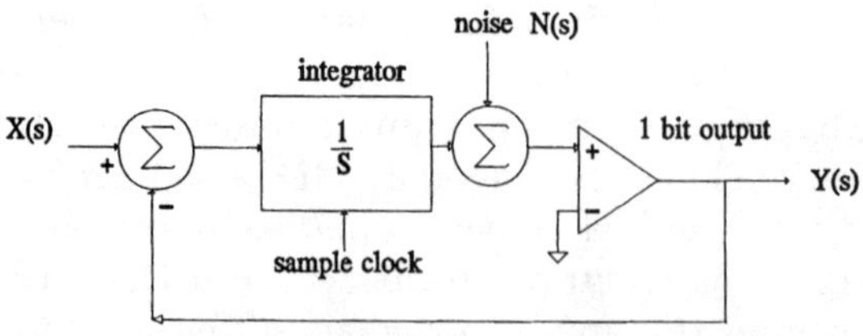

Figure 8.6 Delta-sigma model with noise injection.

$$\frac{Y(s)}{X(s)} = \frac{s}{1+s}$$

Note that the input signal is operated on by a low-pass filter, and the noise is operated on by a high-pass filter. Quantization noise is then pushed out in frequency and is attenuated within the signal band much more than by oversampling alone. This is known as *noise shaping* and is shown in Fig. 8.7. The noise in the upper half of the spectrum can then be filtered out (usually by a low-pass digital filter) for a significant increase in resolution.

What the above analysis indicates is that, with oversampling and the closed-loop modulator, the hardware requirements have been relaxed significantly. Antialias filtering can now be a simple *RC* low-pass filter and in some cases is completely eliminated. Another major advantage is that the quantization can be a simple low-resolution comparator. This has an added benefit of inherent linearity due to just two DAC levels to switch between.

Multiple-Order Modulation

There are many details that distinguish different delta-sigma converters, with the most common being the design of the modulator. Higher-order modulators will attenuate more noise within the signal band, but require more effort to avoid instability. Figure 8.8 illustrates a second-order delta-sigma converter that uses two single stages cascaded together.

The input-output relationship (*Z* transform) is

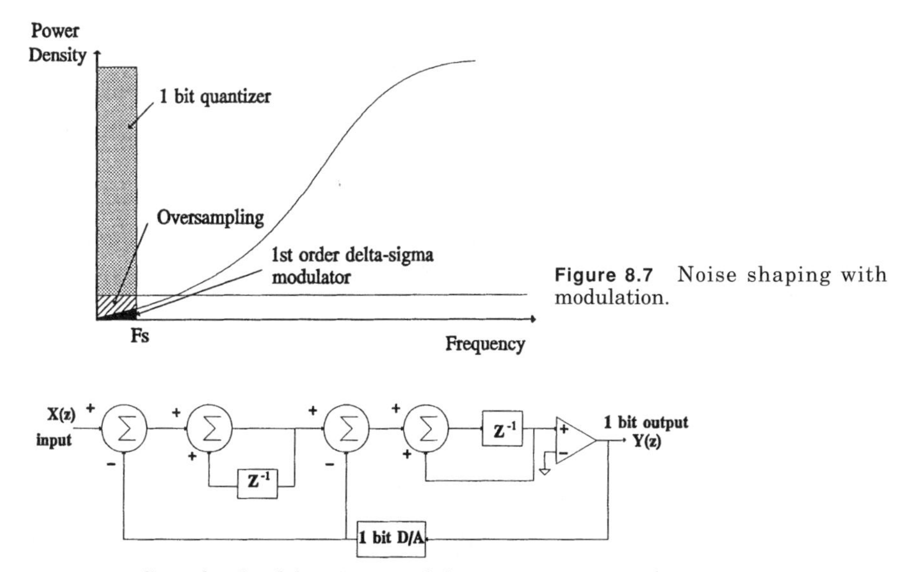

Figure 8.7 Noise shaping with modulation.

Figure 8.8 Second-order delta-sigma modulator.

$$Y(z) = X(z)[z^{-1} + e(1 - z)^2]$$

Note that the input will go straight through to the output after one delay, but the noise will be "shaped" by a second-order high-pass filter. In other words, the noise will be attenuated within the in-band range with only the higher frequencies passed. Adding more modulation stages will increase the order of attenuation. Figure 8.9 shows the consequence of using a higher-order modulator.

Multibit Modulators

Typically, delta-sigma ADCs utilize a simple comparator (a basic two-level DAC) for the benefits of high linearity. However, there are some advantages to increasing the DAC resolution. Multibit quantization decreases the noise exponentially for every bit increase in the modulator ADC. This decreases the noise level by 1 bit for each additional bit in the ADC since the quantization steps (n) are reduced by 2^{-n}. Additionally, the closed-loop feedback is made more stable since the feedback signal can more closely resemble the actual error. This also reduces the requirements of the digital filtering since the out-of-band noise is reduced. Figure 8.10 shows a multibit modulator.

Of course, nothing is for free, and there are some disadvantages to using multibit quantization. First, the ADC must be very fast to set-

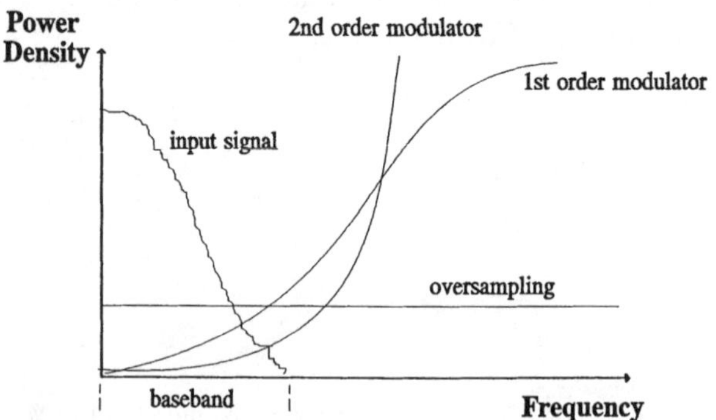

Figure 8.9 Effects of higher-order modulator on in-band signals.

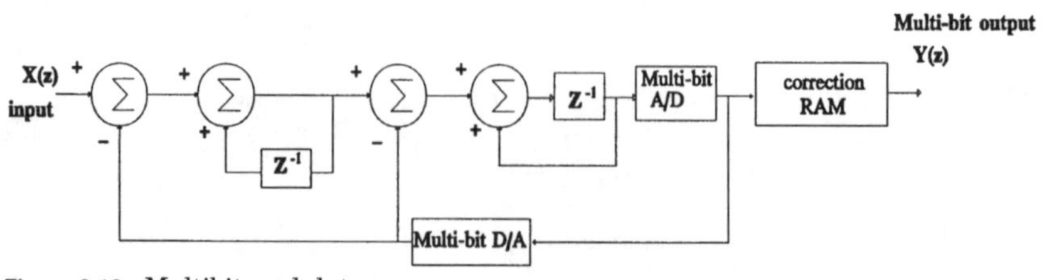

Figure 8.10 Multibit modulator.

tle within the high sampling window. This usually means that a flash converter is called for. The largest concern, though, is the DAC in the feedback loop. This must also be very fast and have high accuracy/linearity to within ½ LSB of the desired output. Any errors in the DAC are directly seen in the output, and some form of compensation will be required. In addition, since the multibit ADC more closely equals the input signal, there will be greater correlation between the input and the DAC error. This will tend to increase the baseband energy content and generate harmonic distortion.

There are several ways to compensate for the problems associated with the multibit modulators. One solution is to use a random circuit to switch in different elements of the DAC to uncorrelate the relationship between the input and output. This will create a white noise effect and will allow the noise level to be filtered out. Other designs utilize an error correction RAM that compensates for the DAC errors such as nonlinearity. This is done by storing the DAC correction codes in RAM during a calibration cycle.

Digital Filters

This section covers the basics of digital filtering as it applies to the delta-sigma method. For much more in-depth treatment of digital filtering (or digital signal processing), the reader should consult the many good references on the topic. Unlike their analog counterparts that operate on a continuous basis, digital filters operate on a sampled-data basis. Figure 8.11 shows a sampled-data representation of a continuous analog input signal. These sampled pulses are operated on by multiplying the amplitudes with a set of coefficients and summing to form a desired impulse response. Virtually any of the classic filter types can be created (i.e., Butterworth, Bessel, Chebyshev) to form a low-pass, bandpass, or high-pass filter. Many other types unique to digital filtering techniques are also possible.

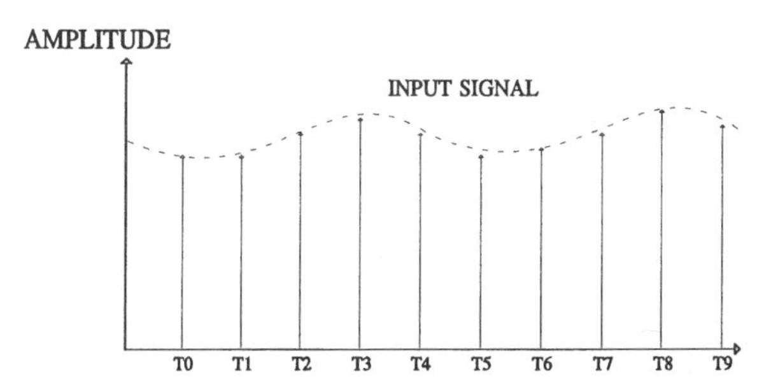

Figure 8.11 Sampled-data representation for digital filtering.

In previous discussions, the focus was on how to create a modulator that could push the quantization noise out in frequency by oversampling and modulation. The idea was to allow the out-of-band noise to be attenuated through high-order digital filtering. Digital filtering plays a major role in making the delta-sigma process practical. Instead of using inductors and capacitors, digital filters use a series of delay elements along with multiplication and addition (accumulate) operations.

Many techniques can be utilized to create a digital filter. In choosing an approach, tradeoffs must be made between complexity and performance (amount of ripple in the passband and amount of group delay). Passband ripple is minimized by increasing the number of filter taps (delay elements). Group delay is the amount of time necessary before an input change can result in a change in the output. This will be proportional to the number of samples taken to complete an output result. The only potentially major problem with using digital filters is the round-off errors from several mathematical operations. This round-off error is reduced by increasing the length of the coefficients. Most delta-sigma A/D converters contain a programmable digital filter that is optimized for a particular application (i.e., instrumentation or audio).

Comparison of digital filters

There are two basic types of filters: the *finite impulse response* (FIR) filter and the *infinite impulse response* (IIR) filter. FIR filters (Fig. 8.12) are also known as *nonrecursive* since an impulse input will eventually cease to cause an output response after a finite number of delays. IIR-type filters (Fig. 8.13), on the other hand, contain feedback terms and are referred to as *recursive.* Theoretically, the response from the IIR filter can continue forever, which is true for any system with feedback. General descriptions of both types of filters follow.

FIR filters

$$y(n) = \sum_{k=0}^{M-1} a(k)x(n-k)$$

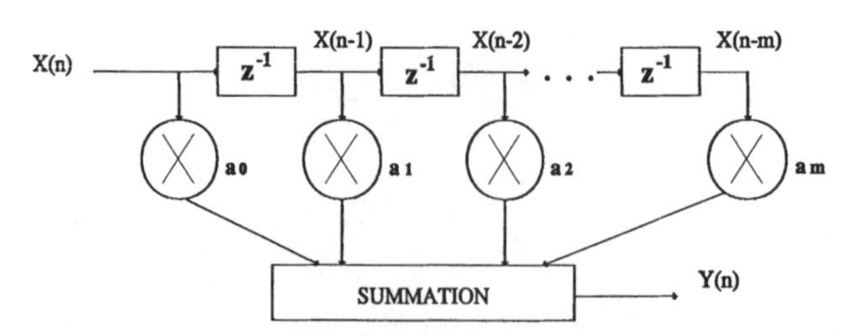

Figure 8.12 Block diagram representation of an *N*th-order FIR filter.

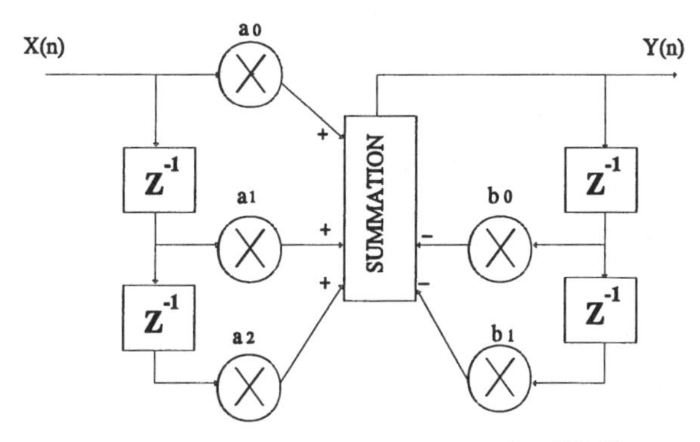

Figure 8.13 Block diagram representation of an IIR filter.

General characteristics of FIR filters:

1. Unconditionally stable since there are no poles (feedback terms).
2. Do not accumulate errors since inputs eventually stop producing a response (no feedback), and thus do not require large coefficients as compared to IIR filters.
3. Easy to design and analyze.
4. May require many more coefficients than IIR filters that will cause lower bandwidth due to the number of multiplications required. This is due to the lack of poles; therefore, roll-off is limited.
5. Can take advantage of decimation for fewer computations per sample.

IIR filters

feedback terms

$$y(n) = \sum_{k=0}^{M-1} a(k)x(n-k) - \sum_{k=0}^{N-1} b(k)y(n-k)$$

General characteristics of IIR filters:

1. They contain feedback elements with poles, and thus higher roll-off is achieved.
2. Computations are more complex than for FIR filters due to the use of historical information of the output required for every sample. Therefore decimation is not possible, and a lower sampling rate must be used.
3. Care must be taken to ensure that the filter will remain stable under all conditions due to the feedback elements.

Decimation

With the input signal oversampled at a very high rate, reducing the final output code frequency will make the computations more manageable. The process of converting short words at high frequency to longer words at a slower rate is known as *decimation*. This greatly reduces the complexity and speed requirements of the mathematical operations. For example, without rate reduction for an N-tap filter, every new output would require N multiplications and additions each sample time. This places a great demand on the processor, especially if the coefficients are several bits long and there are several taps. The term *taps* refers to the number of unit delays utilized in the filter. Note that there are $N - 1$ coefficients for an N-tap filter, as shown in Fig. 8.13.

If the number of computations per sample is reduced, the processor *multiply and accumulate* (MAC) requirements will be relaxed, thus resulting in a lower-cost design and possibly higher performance due to less internal noise from high-frequency digital switching. Often decimation is done simultaneously with the filtering. Usually, several stages are cascaded together with each stage decimating at a different rate. This multirate combination of a filter and decimator structure simplifies the requirements of each successive stage. Shown in Fig. 8.14 is a simple example of how decimation and filtering are accomplished simultaneously. Note that only one multiplication is required for every input sample. Also note that the output rate is reduced by a rate of $1/N$ since a complete result requires the sum of N input samples times N coefficients. The decimation factor in the above example is therefore N.

Delta-Sigma D/A Converters

For systems that require an analog output, the same delta-sigma concepts can be applied. The main difference between delta-sigma ADC and DAC lies in the rate of the output signal. In the ADC section, dec-

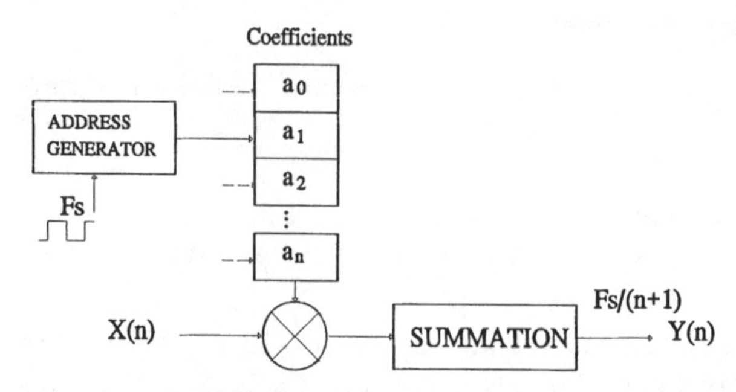

Figure 8.14 An Nth-order FIR filter with $1/N$ decimation.

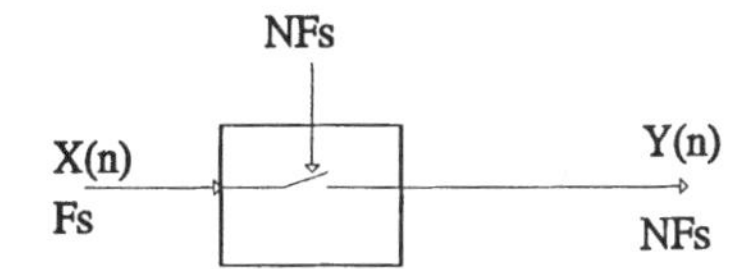

Figure 8.15 Simple interpolation.

imation is used to reduce the high-frequency low-resolution pulses to lower-frequency, higher-resolution words. Delta-sigma DACs, on the other hand, do the reverse. Here a process called *interpolation* is performed that samples the digital outputs at a high rate (see Fig. 8.15). Note that a low-resolution digital code is sampled multiple times at a high rate. This effectively produces a high-resolution/frequency output that is easily low-pass filtered for an analog output.

The 18-bit delta-sigma DAC (CS4328) from Crystal Semiconductor is a good example of a stereo DAC. The CS4328 accepts the digital input from the processor in a serial digital format with the left and right channels controlled by the LRCK input. Note in Fig. 8.16 that there are two channels with an 8 times interpolator and a 64 times oversampling delta modulator for a net oversampling of $8 \times 64 = 512$. The modulator then drives a low-resolution DAC. Following the DAC are a fifth-order switched capacitor and a second-order low-pass filter. The output channels of the CS4328 are then filtered by a 51-Ω resistor and a 0.01-μF capacitor for the final filtering. This provides an analog output with a 120-dB *S*/*N* ratio.

Examples of Delta-Sigma ADC Hardware

There is an ever-growing list of applications for delta-sigma converters. Among the applications are high-performance audio, image processing, communications, and specialized instrumentation. The design of the modulator and digital filter will depend heavily on the application. For example, high-precision dc measurement equipment

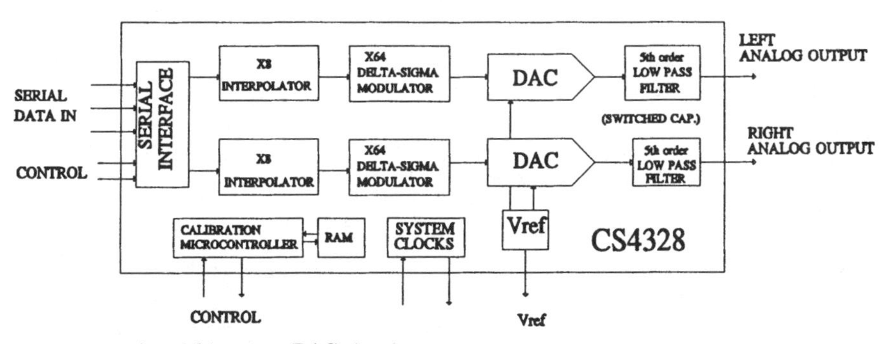

Figure 8.16 An 18-bit stereo DAC circuit.

will typically use a multiple-order filter and/or high sampling rate for high attenuation. Since the signal of interest is rather low in frequency, the long group delays will likely not be an issue. Accuracy, however, is still an issue, and some delta-sigma converters contain the same autocalibration techniques used for successive-approximation ADCs. Users of audio converters, on the other hand, care little about dc accuracy. The only really important specifications are bandwidth and harmonic distortion. Still other delta-sigma converters are aimed at specific applications and contain such features as transducer driver and interface circuitry.

DC measurement by delta-sigma A/D converters

For dc measurements, such as measuring process control signals, delta-sigma A/D converters like the CS5505/6-8 family from Crystal Semiconductor work well. In reviewing the block diagram in Fig. 8.17, note that there are multiple input channels. The on-board multiplexer is available for up to 4 (pseudo) differential inputs with $A_{in}(-)$ serving as a common offset. This is a clue that the converter is aimed at low-frequency signals since there will be some settling time required between channels.

This design has a fourth-order delta modulator followed by a filter made up of a comb filter and a low-pass filter. The comb filter contains 0s that can be placed for maximum line rejection (that is, 50 or 60 Hz). When operated at the recommended frequency of 32 kHz, this filter provides 120-dB attenuation beyond 240 Hz with more than 120-dB rejection at 50, 60, 100, and 200 Hz.

Easy microcontroller interface is provided with the serial communication port. When the DRDY line goes low, the serial output register contains the new result for the microcontroller to serially access the data. This serial port can interface with industry standards such as National Semiconductor's Microwire or Motorola's Serial Peripheral Interface.

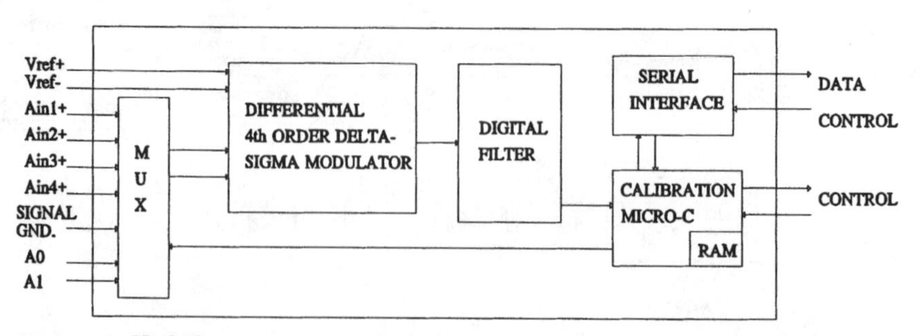

Figure 8.17 High-dc-accuracy 16/20-bit delta-sigma ADC.

Audio delta-sigma A/D converters

There are many reasons why audio engineers frequently turn to digital techniques. DSP is a natural choice for easily improving the quality of recorded material. For instance, a signal-to-noise ratio *S/N* of greater than 16 bits can be achieved. This would be very difficult with other techniques (i.e., successive-approximation methods). To perform equalization, all that is required is the use of a set of digital filters with corresponding gain coefficients for different frequency bands. Other advantages of digital techniques include mixing without added noise and reverb using digital delays for concert hall performance.

An example application using the delta-sigma ADC for digital audio is the Crystal Semiconductor CS5326-29 family. These devices provide 16- and 18-bit stereo performance. The block diagram in Fig. 8.18 shows two separate input channels, each with its own differential input, delta modulator, and digital filter. Differential input stages assist in providing low-*S/N* operation. The modulator runs at a 64 times oversampling rate, and the digital filters consist of a three-stage filter.

Serial outputs at a rate of up to 50 kHz and either 16 or 18 bits are alternately read by toggling the L/R input pin. Usually, the outputs are first read by a digital signal processor for further processing and storage before they go on to the DAC output stages. Note that there is a small microcontroller with calibration RAM resident within the ADC. This provides offset correction of the input stages after initial power-up. This prevents the possibility of hearing the crackling sound after the equipment is turned on.

Specialized delta-sigma A/D converters

An example of a very flexible delta-sigma A/D converter designed for instrumentation applications is the Analog Devices AD7710/11. As shown in Fig. 8.19, there are several additional circuits that ease

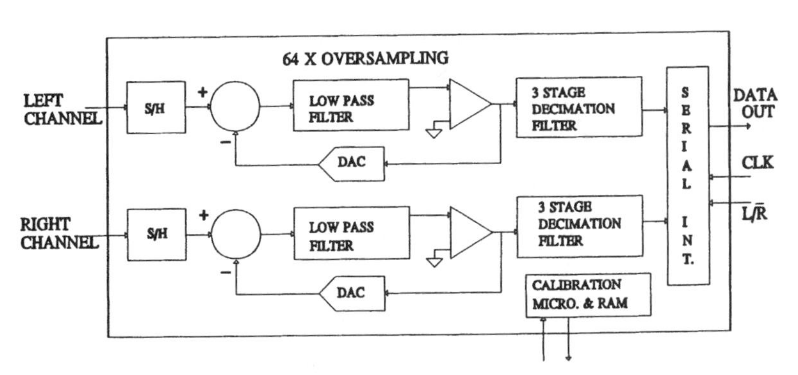

Figure 8.18 Block diagram of CS5326-9 digital audio ADC.

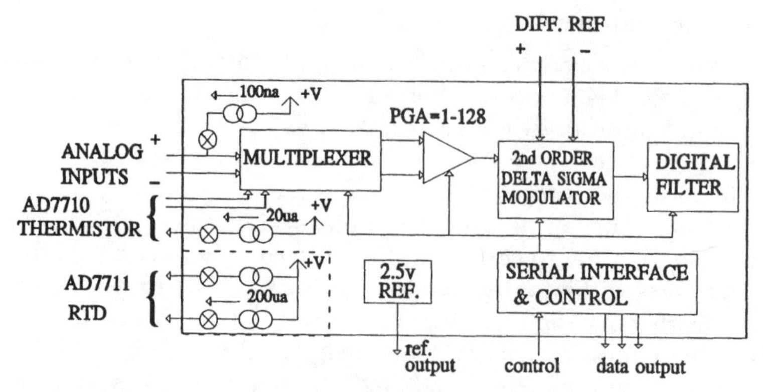

Figure 8.19 Analog devices AD7710/11 instrumentation delta-sigma ADC.

interfacing to a thermocouple, a *resistance temperature detector* (RTD), or a strain gage. For example, the 100-nA switchable current source will detect an open thermocouple wire. When temperature is measured with a thermocouple, it is necessary to also measure the temperature of the circuit board. This is called *cold junction compensation,* and it refers to the temperature offset voltage effects of the thermocouple wires soldered into the circuit board. By connecting a thermistor to the 20-μA current source and reading the analog voltage along with the thermocouple, the system microcontroller can compensate for the temperature changes.

An RTD is capable of measuring very wide temperature ranges (that is, 800°C). The AD7711 provides a 200-μA current source that transforms the RTD change versus temperature to voltage for the ADC. When an RTD is used, compensation is required to deal with the nonlinearity of the RTD sensor. Luckily, this error versus temperature is a well-defined parabolic curve, and it can be taken care of with the system microcontroller.

Strain gages can also be measured with the AD7710/11. The differential reference is used for the ADC as well as the bridge driver, and the differential input channel measures the bridge as described in Chap. 6 (dual-slope ADC example).

Other features of the AD7710/11 include a *programmable gain amplifier* (PGA), 2.5-V reference, delta-sigma ADC, and easy serial interface. The PGA following the multiplexer provides added flexibility for measuring wide input signal ranges. Resolution up to 20 bits is achieved with the second-order delta modulator operating at a very high oversampling rate and an on-board digital filter.

Chapter

9

System Error Analysis and Control

There are plenty of potential error sources lurking outside the A/D converter, and if they are not taken into account, they can degrade the performance of any design. These error sources can be either internally generated from components within the data acquisition system or externally generated (i.e., from power supplies, radio-frequency interference). This chapter focuses on minimizing all unwanted signals from the system including dc tolerance stack-ups, ac noise, and temperature/time drifts. All the support circuits including amplifiers and buffers, references, and discrete resistor components produce some magnitude of these errors. Quite often, the resistor network used for scaling and gain plays a vital role in the final system errors (initial and drift). In addition to support components, problems can result from printed-circuit board leakage, ground loops, and high-frequency ac signal coupling.

As analog system accuracies increase, so does the importance of understanding the limitations of each component. In many high-accuracy designs (i.e., greater than 12 bits), it is impractical to eliminate the system reference, offset voltages, and gain errors with hardware alone. Frequently, sensors such as strain gages and thermocouples initially pose tolerances that make calibration a necessity. In these applications, initial component tolerances are not important. Once the system has been calibrated, the design only needs to maintain low drift specifications. Software-driven autocalibration is usually preferred to a totally manual approach and can be done with an operator interface on a periodic basis.

DC Tolerance Stack-Ups

Before a data acquisition (or any precision analog circuit) design is started, the total accuracy target must be known. This really is a budget for errors that will be distributed among the entire system components. The costs associated with achieving the desired individual component tolerances usually factor into the decision on where to spend your budget. Which error sources are most important depends on the application requirements and usually requires careful analysis of the manufacturer's data sheets. The beginning sections describe the dc error sources and compensation techniques related to each of the support circuits common in a data acquisition system. After a review of the resulting error terms, an optimum solution can likely be reached by focusing on the largest contributor.

Amplifier errors

Data acquisition systems often require an amplifier to scale, offset, and buffer an input signal from a sensor. There are several limitations associated with the real-world operational amplifier (op amp). In addition to amplifier errors, the resistors used for amplification must be considered (discussed in Chap. 2). The op amp dc errors are listed below.

Op amp dc errors:

1. Input offset voltage V_{os} and drift
2. Input bias offset current I_{bos} and drift
3. Limited open-loop gain (A_v)
4. Common-mode rejection ratio (CMRR)

Voltage offsets

The differential amplifier in Fig. 9.1 illustrates the effects of the above-mentioned error sources. Bipolar op amps have input offset voltages caused by mismatches in the differential input stage V_{be} values (base emitter voltages). The output offset error V_{oe} produced by the input offset V_{os} is computed from

$$\text{Output offset error } V_{oe} = \pm V_{os}\left(1 + \frac{R_1}{R_2}\right)$$

The above equation is true for both an inverting and a noninverting amplifier circuit. This is because the voltage offset is relative to both inputs; therefore the larger noninverting gain is used. Another poten-

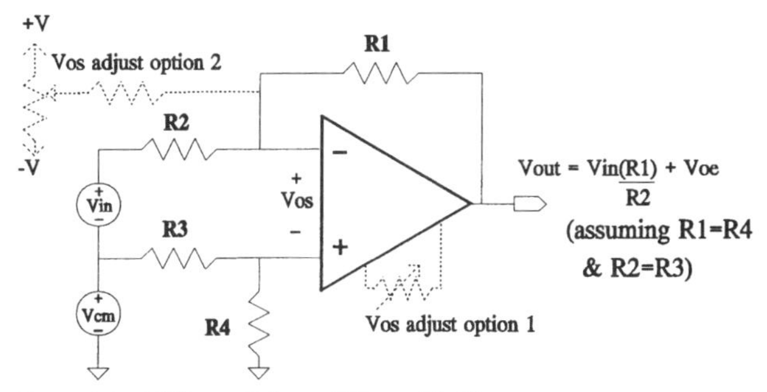

Figure 9.1 Differential amplifier with V_{os} error.

tially more severe problem is the variation of input offset voltage with time and temperature. Initial offsets can be easily calibrated out, while drifts are considerably more difficult since full characterization is required over the operating temperature range.

There are numerous op amps available with offset voltages below 100 μV that can eliminate concern with V_{os} in many applications. Some amplifiers with higher V_{os} values provide a trim adjustment (option 1 in Fig. 9.1) which alters the current through the op amp input transistor stage. However, this imbalance in current levels will degrade the temperature drift performance, so only a slight adjustment is recommended. If necessary, a resistor circuit with a potentiometer connected to a positive and negative supply (option 2) can be used to zero out the error, although most high-volume products cannot tolerate the cost of a manual adjustment, unless error sources other than the amplifier demand it.

Current offsets

Bipolar op amps require some input bias current I_b to operate due to the front-end base connections. The direction of the current is out of the input pins for a PNP input stage and into the input pins for an NPN stage. The potential problem with I_b will vary depending on the external circuit resistances and amplifier selection. As an example, Fig. 9.2 shows a noninverting amplifier with bias currents I_{b1} and I_{b2} causing an output voltage error V_{oe}. Errors are calculated by

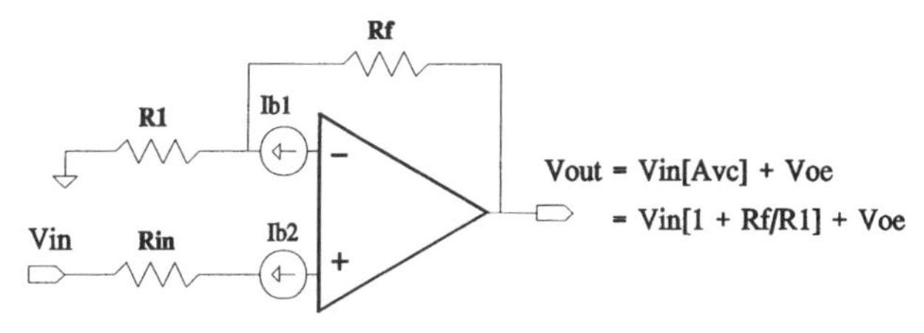

Figure 9.2 Example of I_b and I_{bos} output errors.

first determining the equivalent resistance seen by the amplifier inputs and then multiplying them by the individual bias currents.

$$V_{oe} = A_{vc}(I_{b2}R_{in} - I_{b1}R_{eq}) \qquad R_{eq} = R_f || R_1$$

For example, assuming that $I_{b1} = I_{b2} = 250$ nA and

$$R_{in} = 10\text{k}\Omega \qquad R_1 = 1\text{ k}\Omega \qquad R_f = 99\text{ k}\Omega$$

Therefore

$$A_{vc} = 1 + \frac{99}{1} = 100$$

$$V_{oe} = 100\ (250\text{ nA})\left[10\text{ k}\Omega - \frac{1\text{ k}\Omega(99\text{ k}\Omega)}{100\text{ k}\Omega}\right]$$

$$= 225\text{ mV}$$

Output offset errors from I_b can be easily eliminated by setting the equivalent input resistances seen by the positive and negative inputs the same (that is, $R_{in} = 1$ kΩ in the previous example), though there will always be a difference between the input currents I_{b1} and I_{b2} that will leave a portion of the error. This difference in input bias currents is called the *input bias offset current* $I_{bos:}$

$$I_{bos} = \pm\ (I_{b1} - I_{b2})$$

Typically, I_{bos} will be much less than $I_{b.}$ For example, the LM324 op amp has an I_b of 250 nA and I_{bos} of only 50 nA. Therefore, the output error with $R_{in} = R_{eq}$ in the previous example will be

$$V_{oe} = A_{vc}R_{eq}I_{bos} = 100\ (1\text{ k}\Omega)(50\text{ nA}) = 5\text{ mV}$$

It is obvious that low-value resistors should be used (i.e., less than 1 kΩ), and equivalent input resistances should be the same to minimize the error from input bias currents. Input bias offset currents and voltage offsets have the same net effect on the output; hence it is possible to use the same compensation (options 1 and 2 in Fig. 9.1). If manual calibration is undesirable due to cost or temperature degradation, then another amplifier must be chosen that has the required I_{bos} and V_{os} values. For example, a CMOS amplifier has extremely low input current compared to a bipolar; therefore it can tolerate a much higher input resistance.

Limited open-loop gain

Ideally, the op amp provides infinite gain, but of course it is not ideal, and there is an error associated with it. Typically, open-loop gain A_v greater than 100 k at direct current will likely not cause a problem with low- to moderate-accuracy designs. When you review an amplifier data sheet, the open-loop gain is expressed in volts per millivolt or decibels. High-precision designs will be affected if the gain is insufficient, especially when the input signal is no longer at direct current. This is due to the amplifier gain roll-off versus frequency. To see the effect on the closed-loop gain by a finite open-loop gain, look at the true closed-loop gain equation for a noninverting amplifier.

Amplifier closed-loop gain equation:

$$A_{vc} = \frac{A_v}{1 + B_{Av}} \text{ (noninverting gain circuit)}$$

where A_{vc} = amplifier closed-loop gain
A_v = amplifier open-loop gain
$B = R_1/(R_1 + R_f)$, R_f = feedback, R_1 = input

As an example, suppose an amplifier is used in the circuit of Fig. 9.2 with the following conditions:

$$A_v = 100 \text{ V/mV or } 100 \text{ k} \qquad \text{(at direct current)}$$

$$A_{vc} = 100 \qquad \text{for dc signal}$$

Determine the percentage of gain error:

$$B = \frac{1}{1 + 99} = 0.01$$

$$A_{vc} = \frac{100 \text{ k}\Omega}{1 + 0.01\,(100 \text{ k}\Omega)} = 99.93$$

Therefore,

$$\text{Gain error} = \frac{100\,(100 - 99.93)}{100} = 0.07\%$$

Note: This does not include the gain errors from the circuit resistors, which will vary over temperature. In addition, as the input signal frequency increases, the open-loop gain will decrease [i.e., 20 dB/decade, causing more gain error.

The dependency of the gain error on the open-loop gain is made more obvious with the following example calculations. Solving the closed-loop gain equation for the required open-loop gain, given A_{vc} and the desired accuracy, yields

$$A_v = \frac{A_{vc}'}{[1 - BA_{vc}']}$$

where A_{vc}' = attenuated closed-loop gain
= $(1/B)(1 - \%\ \text{error}/100)$

	Desired accuracy			
	0.4% 8-bit	0.1% 10-bit	0.025% 12-bit	0.0015% 16-bit
Gain = 10, A_v =	2.5 k	10 k	40 k	660 k
Gain = 100, A_v =	25 k	100 k	400 k	6.6 M

Note that the required open-loop gain reaches a tremendous value as the closed-loop gain and/or desired accuracy is increased.

Possible solutions to improve the error caused by limited open-loop gain include these:

1. Select an op amp with sufficient gain over the desired operating frequency range.
2. Minimize the circuit closed-loop gain, or use two amplifiers in cascade if necessary (that is, $A_{vc1} = 10$ times $A_{vc2} = 10$ for a total $A_{vc} = 100$).
3. Adjust out the gain error, and keep the temperature relatively constant near the op amp if necessary.

Common-mode error

Revisiting the differential circuit in Fig. 9.1, we note that both inputs are biased up by a common voltage (V_{cm}). Due to the nonideal nature of every op amp, there will be some output error directly proportional to V_{cm} and inversely proportional to V_{in} and the common-mode rejection ratio. In other words, as the input signal decreases and the common-mode voltage increases, the common-mode rejection becomes more important. The result is that some finite amount of the common-mode voltage will appear at the amplifier output along with the desired signal. This error can be determined by using the common-mode rejection ratio (CMRR) in the data sheet for a particular op amp. The CMRR is defined as

$$\text{CMRR} = \frac{A_v}{A_{vcm}}$$

where A_v = differential gain
A_{vcm} = common-mode gain
$= V_{out}/V_{cm}$

Example of common-mode error

Determine the output error V_{oe}, given a CMRR of about 100 dB (or 100 k) at direct current, a V_{cm} of 10 V, a differential gain A_v of 10, and V_{in} = 1 V. Rearranging the above equation, we have

$$A_{vcm} = \frac{A_v}{\text{CMRR}}$$

$$= \frac{10}{100\text{ dB}} = \frac{1}{10\text{ k}}$$

Therefore,

$$V_{oe} = \frac{V_{cm}}{A_{vcm}} = \frac{10\ V}{10\text{ k}} = 1\text{ mV}$$

$$\%\text{ error} = \frac{100\ V_{oe}}{V_{in}A_v}$$

$$= \frac{100 \times 1\text{ mV}}{1\text{ V} \times 10} = 0.01$$

Keep in mind that the above result does not include the mismatches in the resistor network which tend to dominate the total CMRR error. Suppose we used the differential amplifier example in Fig. 9.1 with the above conditions and a 0.1 percent mismatch in network resistance. What would the percentage of output error be? Assume a unity-gain-differential circuit with worst-case values for R_2 and R_4 of + 0.1 percent and for R_1 and R_3 of − 0.1 percent. This causes a net 0.2 percent error at the output. Therefore the common-mode error equals

$$V_{oe} = V_{cm}\text{ (in)} \times 0.2\% = 10\text{ V} \times 0.2\% = 20\text{ mV}$$

$$\%\text{ output error} = 100 \times \frac{20\text{ mV}}{1\text{ V}} = 2$$

Note that there is a multiplier effect of the percentage of mismatch in circuit resistances times V_{cm}. Even though 0.1 percent resistors are used, there is a much greater output error (2 percent) that is caused by the common-mode voltage of 10 V. Also, the common-mode error will increase with poor temperature-tracking resistors. As discussed in Chap. 2, thin-film resistor networks offer the best temperature-tracking characteristics.

Reference errors

Reference tolerance for an absolute measurement will directly affect the system accuracy. For instance, a 0.1 percent reference tolerance causes a 0.1 percent output error at full-scale input. If the measurement is ratiometric (i.e., the transducer signal is proportional to the reference signal), then there is no dc error contribution. Absolute reference errors show up in the final measurement as gain error, and they can be compensated for in a number of ways. For example, an autogain technique is discussed at the end of this chapter, or with an amplifier manual gain adjustment will eliminate the reference error.

The temperature/time drift and the output noise are usually more important than the reference initial tolerance for absolute measurements. This is particularly true when the system sensor and other components such as op amp circuits do not meet the desired accuracy. This will make calibration of the system reference voltage mandatory. Even with calibration, reference noise can still be an issue due to the much higher noise generated in comparison to an amplifier. (*Note:* refer to Chap. 3 for a more detailed discussion on highly stable and low-noise reference circuits.)

AC Noise Sources

System noise can result from components such as resistors, references, and amplifiers used to condition the signal for a data acquisition system. Other potential noise sources include high-speed digital circuitry, power supplies, and external sources (i.e., RFI, EMI). Recommended layout practices covered later in this chapter are ways to limit these unwanted signals. The focus of this section is on managing component noise. Transistors and resistors inherently generate voltage, current, and thermal noise. This somewhat erratic behavior of the current flow is a natural phenomenon due to the properties of the devices. Whether the generated noise signals are a real problem will depend on the magnitude of the input signal. With the root-mean-square (rms) noise typically in the microvolt range, only very low input signals (i.e., strain gage sensors) and/or high-accuracy designs are affected.

An important point is that the input signal must be sufficiently greater than the noise at the front-end amplifier stage. It is here that the “noise floor” is established which determines the smallest input

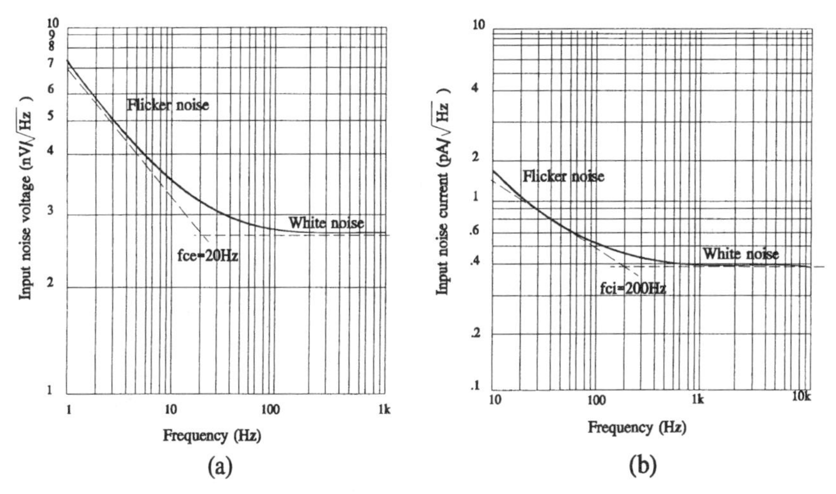

Figure 9.3 Spectral noise density of the LM627 op amp. (*a*) Voltage noise density; (*b*) current noise density.

signal measurement possible. Once the signal is amplified, the remaining component noise levels are not as critical.

Amplifier ac noise

We use an example to understand the characteristics of voltage, current, and thermal noise present in an amplifier. Figure 9.3 shows the voltage and current spectral noise density plots for a typical low-noise amplifier. It is customary to specify the noise density in units of $\mathrm{pA}/\sqrt{\mathrm{Hz}}$ for current and $\mathrm{nV}/\sqrt{\mathrm{Hz}}$ for voltage versus frequency on a logarithmic scale. Note that there are two portions to each curve. The lower-frequency portion (less than 100 Hz for a low-noise amplifier) indicates that there is an inverse relationship between density and frequency. This is referred to as either $1/f$ or flicker noise and is present in all conducting devices. The second portion is flat with respect to frequency, and therefore the noise density is constant over the entire frequency range. This is referred to as *white noise* since it contains a wide range of frequencies, which is analogous to white light containing several wavelengths. The intersection of flicker and white noise density slopes is called the *corner frequency* (f_{ci} for current, and f_{ce} for voltage), and as you will see in the following discussion, it is very important.

Before we try to calculate the noise produced by an amplifier circuit, first let's define the variables used in the op amp noise model of Fig. 9.4:

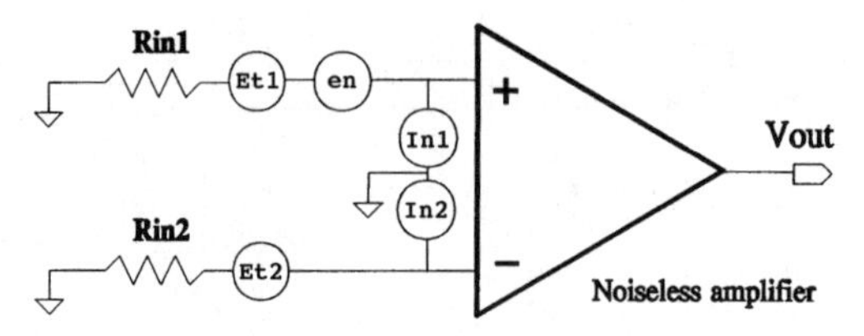

Figure 9.4 Noise model for an op amp.

E_n = total equivalent input voltage noise source

$I_{n1,2}$ = equivalent input current for noise sources 1 and 2

$E_{t1,2}$ = resistor thermal noise sources 1 and 2 (explained later)

Equations for predicting the input referred noise can be derived from the spectral noise density plots (Fig. 9.3). The total input referred noise is arrived at by combining both white noise and flicker noise contributions for a given frequency band ($f_H - f_L$). This relates the noise density to an rms value.

White noise portion:

$$E_{n(\mathrm{wh})} = e_{n(\mathrm{wh})}\sqrt{f_H - f_L}$$

$$I_{n(\mathrm{wh})} = i_{n(\mathrm{wh})}\sqrt{f_H - f_L}$$

Combining the flicker and white noise is a little more complicated and is accomplished by using the corner frequency breakpoints (f_{ce} and f_{ci}).

Flicker noise portion based on white noise and corner frequencies:

$$e_{n(\mathrm{fl})} = e_{n(\mathrm{wh})}\sqrt{\frac{f_{ce}}{f}} \qquad f_{ce} = \text{voltage corner frequency}$$

$$i_{n(\mathrm{fl})} = i_{n(\mathrm{wh})}\sqrt{\frac{f_{ci}}{f}} \qquad f_{ci} = \text{current corner frequency}$$

Therefore, the flicker voltage noise equals

$$E_n^{\,2} = \int_{f_L}^{f_H} \left[e_{n(\mathrm{wh})}\sqrt{\frac{f_c}{f}} \right]^2 df$$

$$= e_{n(\text{wh})}^{2} f_{ce} \ln \frac{f_H}{f_L}$$

Similarly, the flicker current noise equals

$$I_n^{2} = i_{n(\text{wh})}^{2} f_{ci} \ln \frac{f_H}{f_L}$$

From the above, the total voltage and current noise must be computed by using both flicker and white noise contributions.

$$E_n = e_{n(\text{wh})} \sqrt{f_{ce} \ln \frac{f_H}{f_L} + f_H - f_L}$$

$$I_n = i_{n(\text{wh})} \sqrt{f_{ci} \ln \frac{f_H}{f_L} + f_H - f_L}$$

Before the total input referred noise can be computed, it is necessary to define one more term. The resistor thermal noise E_t, also known as the *Johnson noise,* is a white noise voltage generated from thermally charged carriers within resistors. This noise is present in discrete resistors as well as in the amplifier front end. However, the amplifier voltage noise specification accounts for this. Below is the thermal noise equation. Thermal noise:

$$E_t = \sqrt{4kTR(f_H - f_L)}$$

where k = Boltzmann constant of 1.38×10^{-23} J/K
T = absolute temperature, K
R = circuit resistance, Ω

At room temperature, E_t simplifies to

$$E_t = 1.28 \times 10^{-10} \sqrt{R_x(f_H - f_L)}$$

Finally, the total input referred voltage noise source can be arrived at by simply taking the rms of each term above. To do this, it is necessary to convert the current term $[I_n(f_H, f_L)]$ to voltage by multiplying it by the input resistances (R_{in1}, R_{in2}). Therefore, the total input referred noise in terms of voltage noise is described by

$$E_n(f_H, f_L) = \sqrt{E_n^{2} + (I_n R_{\text{in1}})^2 + (I_n R_{\text{in2}})^2 + E_{t1}^{2} + E_{t2}^{2}}$$

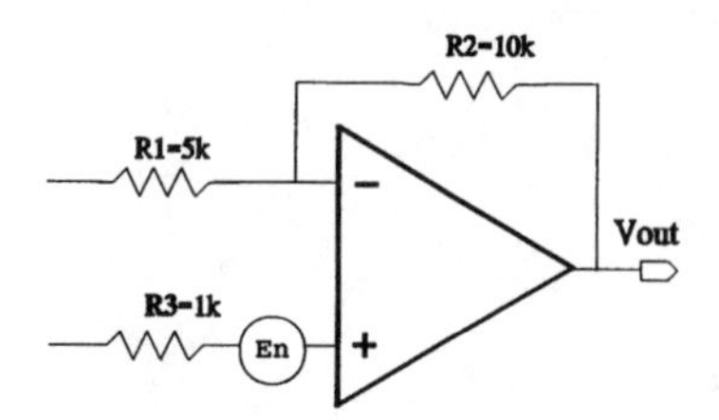

Figure 9.5 Example of input referred voltage noise.

Application examples

Using a design example will help make the above theory much clearer. Figure 9.5 shows an LM627 amplifier with the frequency range limited to 0.0001 Hz to 1 kHz. In the following example, we determine the total input referred noise by using the noise model in Fig. 9.4 and the above equation.

Example 9.1 LM627

First, compute I_n.

$$I_n = i_{n(\text{wh})}\sqrt{f_{ci}\ln\frac{f_H}{f_L} + f_H - f_L}$$

$$= (0.4\ \text{pA})\sqrt{200\ln\frac{1\ \text{k}}{0.0001} + 1\ \text{k} - 0.0001}$$

$$= 22\ \text{pA rms}$$

Second, compute the thermal noise (E_{t1} and E_{t2}).

$$R_{\text{in1}} = R_1 || R_2 = 3.33\ \text{k} \qquad R_{\text{in2}} = R_3 = 1\ \text{k}$$

$$E_t = 1.28 \times 10^{\nabla-10}\sqrt{R_x(f_H - f_L)}$$

$$E_{t1} = 1.28 \times 10^{-10}\sqrt{3.33\ \text{k} \times 1\ \text{k}} = 0.23\ \mu\text{V rms}$$

$$E_{t2} = 1.28 \times 10^{-10}\sqrt{1\ \text{k} \times 1\ \text{k}} = 0.13\ \mu\text{V rms}$$

Fourth, compute E_n.

$$E_n = e_{n(\text{wh})}\sqrt{f_{ce}\ln\frac{f_H}{f_L} + f_H - f_L}$$

$$= 3\text{ nV}\sqrt{20 \ln \frac{1\text{ k}}{0.0001} + 1\text{ k}}$$

$$= 0.11\ \mu\text{V rms}$$

Fifth, compute the total input referred noise voltage.

$$E_n(f_H, f_L) = \left[E_n{}^2 + (I_n R_{\text{in}1})^2 + (I_n R_{\text{in}2})^2 + E_{t1}{}^2 + E_{t2}{}^2\right]^{1/2}$$

$$= \Big[0.11\text{ nV}^2 + [22\text{ pA}\,(3.33\text{ k})]^2 + [22\text{ pA}\,(1\text{ k})]^2 + 0.23\ \mu\text{V}^2$$

$$+ 0.13\ \mu\text{V}^2\Big]^{1/2}$$

$$= 0.28\ \mu\text{V rms}$$

For the peak-to-peak voltage, multiply the rms value by 6. This provides a probability of about 99.7 percent that the input referred noise will be less than that calculated.

$$E_{n\text{p-p}} = 6 \times 0.28\ \mu\text{V rms} = 1.65\ \mu\text{V}_{\text{p-p}}$$

Keep in mind that component noise is very random and that the calculated value is what you would expect to measure over a long time. Therefore, statistics are used to predict the noise, not to define it exactly.

Since the calculated noise is input referred noise, it should be compared to the input signal to determine if there is adequate S/N. If the noise is unacceptable, there are some possible reduction approaches. Close examination of the above calculation reveals that thermal noise dominates. If the equivalent input resistances and/or bandwidth can be reduced, the noise in this example will be lowered significantly. Be careful, though, not to lower the resistance to the point that self-heating becomes a resistance-temperature coefficient problem. Another way to reduce the noise is to reduce the bandwidth, as shown in the shaded areas of Fig. 9.6. By using the RC network, a low-pass filter is created that attenuates the noise as a result of f_H being reduced.

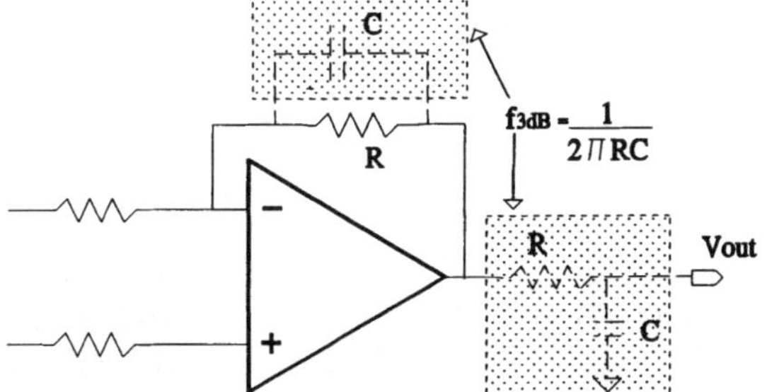

Figure 9.6 Options for bandwidth reduction.

As a second example, the National Semiconductor LMC660 CMOS amplifier is used with the above set of conditions. This amplifier has extremely low input leakage current; consequently the current noise density of 0.0002 pA/Hz$^{1/2}$ is negligible. However, the voltage noise density and corner frequency are much higher than those for the bipolar amplifier in Example 9.1.

Example 9.2 LMC660 CMOS amplifier

$$f_{ce} = 200 \text{ Hz} \qquad e_{n(\text{wh})} = 20 \text{ nV/Hz}^{1/2}$$

First, compute E_n.

$$E_n = e_{n(\text{wh})}\left(f_{ce} \ln \frac{f_H}{f_L} + f_H - f_L\right)^{1/2}$$

$$= 20 \text{ nV}\left(200 \ln \frac{1 \text{ k}}{0.0001} + 1 \text{ k}\right)^{1/2}$$

$$= 1.3 \text{ µV rms}$$

Second, compute the total input referred noise voltage, using the same equation as in Example 9.1 including the resistor thermal noise.

$$E_n = \left(1.3 \text{ µV}^2 + 0.22 \text{ µV}^2 + 0.13 \text{ µV}^2\right)^{1/2}$$

$$= 1.3 \text{ µV rms} \qquad \text{or} \qquad 6 \times 1.3 \text{ µV} = 7.8 \text{ µV}_{\text{p-p}}$$

As you can see from the above analysis, the voltage noise in the CMOS amplifier is by far the dominant factor.

Popcorn noise

There is another "interesting" source of amplifier noise caused by defects in the manufacturing processing that was not included in the previous analysis. *Popcorn noise* is very random pulselike noise that gets its name from the sound it can make through a loudspeaker. All of a sudden, the white noise level will jump for a few milliseconds and then vanish. Because of this extremely hit-or-miss nature, popcorn noise is very difficult to predict or measure. Fortunately, manufacturers have become quite good at preventing popcorn noise because the test time to screen it out would be cost-prohibitive.

Power supply rejection ratio (PSRR)

One more source of noise error associated with an amplifier that requires attention is the input referred noise caused by limited rejection of power supply ripple. The *power supply rejection ratio* (PSRR) is defined as the change in input referred offset voltage for a given change in power supply voltage. Op amp data sheets usually provide a plot showing PSRR in decibels or microvolts per volt versus frequency. As an example, the LM627 has about 140-dB PSRR at direct current with a 20-dB roll-off for both the positive and negative power supply inputs. If a noninverting LM627 amplifier circuit with a gain of 100 is used, and if the power supply has 10-mV rms ripple at 10 kHz, determine the output noise.

Example 9.3 PSRR error

First, determine the PSRR at 10 kHz.

$$\text{PSRR} = 140 - 20 \log (10\text{ k}) = 60 \text{ dB}$$

Second, determine the input referred offset voltage.

$$V_{os} = \frac{0.01 \text{ V}}{60 \text{ dB}} = \frac{0.01}{1000}$$
$$= 0.01 \text{ mV rms}$$

Third, determine V_{oe}; multiply by $1 + R_f/R_{in}$.

$$V_{oe} = (0.01 \text{ mV})(1 + 100) = 1 \text{ mV rms}$$

Whether or not the above is an acceptable error will depend largely on the actual input signal level (or its gained output) and the desired S/N ratio. In many cases, the power supply noise can be the largest source of error. This is especially true if a switching supply is used, but it can also be a problem in a linear supply that is poorly regulated or filtered. When possible, it is preferable to minimize the power supply noise at its source rather than to cover it up elsewhere in the design. If the power supply noise needs to be attenuated, the circuit in Fig. 9.7 is a typical solution. When an RC filter such as this one is used, it is important to pay attention to the amplifier output current requirement. Voltage drops caused by the filter resistors can potentially cause more output errors than the power supply ripple voltage did initially.

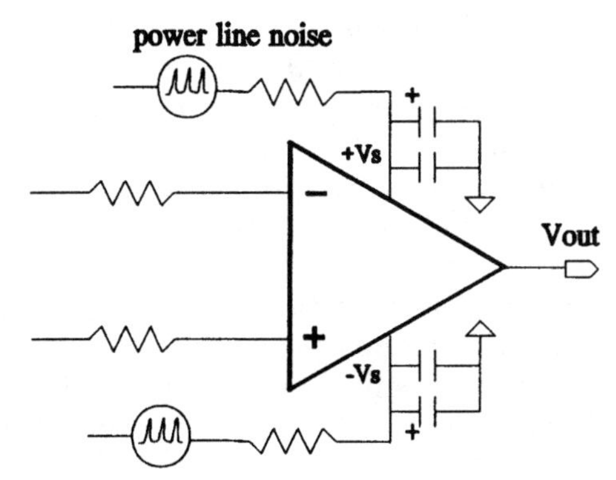

Figure 9.7 Example of filtering for power supply noise reduction.

Low-noise amplifier design guidelines

For very low-noise amplifier designs, consider the following summary of recommendations:

1. Select an op amp with low white noise levels as well as low corner frequencies.
2. Use as low as possible input resistor values while avoiding self-heating.
3. Limit the amplifier bandwidth to match the desired input signal bandwidth. Use a low-pass filter for near dc signals. Use the band-pass filter for ac coupled signals to eliminate flicker noise and attenuate white noise.
4. Use a well-regulated linear supply and a well-designed power supply filter.
5. Use careful circuit board layout practices (discussed later in this chapter).

Temperature Drifts

When the dc tolerance can be controlled or calibrated out and the noise is in an acceptable range, the temperature drift remains the last obstacle to overcome. Temperature drift for high-accuracy designs is a major problem that can be a real challenge and/or costly to eliminate. Besides the amplifier, reference, and resistance temperature characteristics, the circuit board thermocouple effect can be troublesome.

Amplifier temperature drifts

There are primarily two specifications pertaining to the op amp that are sensitive to temperature changes, input offset voltage, and input bias current offset. Input offset voltage drifts versus temperature for amplifiers range from about 20 µV/°C for the low-end versions to as low as 0.2 µV/°C for precision amplifiers. Figure 9.8 shows some typical input offsets as a function of temperature for the LM627 amplifier. Keep in mind that these plots are typical of only a small sample and cannot be a guarantee for every amplifier produced. Note that the individual sample devices vary considerably in both positive and negative directions. The magnitude of the V_{os} and I_{bos} temperature drifts can also approach the worst-case specifications. Therefore, wide temperature changes will negate calibrations made for V_{os} and I_{bos}. Extra effort to provide temperature compensation is rather difficult since the errors are nearly impossible to predict. Preferably, the

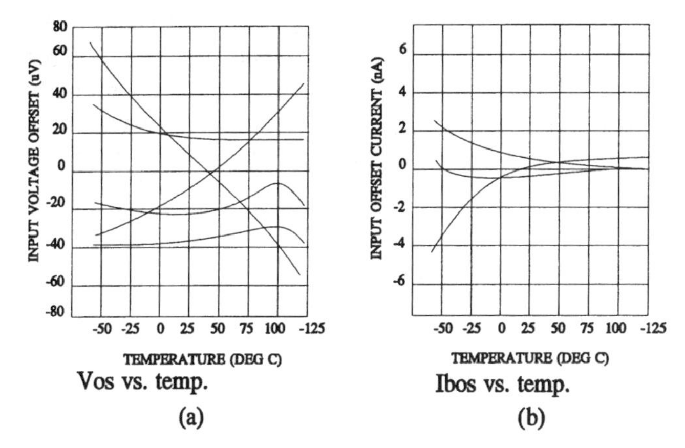

Figure 9.8 Typical LM627 V_{os} and I_{bos} temperature drifts: (*a*) V_{os} versus temperature; (*b*) I_{bos} versus temperature.

errors should be minimized from the start by choosing an amplifier that meets the application requirements.

Errors caused by the V_{os} temperature drift are straightforward to calculate. Once the temperature range is factored in, the worst-case V_{os} change can be calculated easily. For example, a 70°C change with the LM627 amplifier having 0.3 μV/°C temperature drift will produce the following worst-case error:

$$\Delta V_{os}(t) = 70 \times 0.3\ \mu\text{V} = 21\ \mu\text{V}$$

Worst-case input bias offset current over the maximum operating temperature range is always stated in the manufacturer's data sheets (that is, **nA**). The typical plots provide only some indication of what to expect; therefore the designer is forced to use the worst-case specification provided (regardless of the actual temperature change). For instance, the LM627 has a maximum I_{bos} of 20 nA over temperature. The output error with a feedback resistance of 20 kΩ in a voltage-follower configuration will therefore be

$$\Delta V_{out}(t) = (20\ \text{nA})(20\ \text{k}\Omega) = 0.4\ \text{mV}$$

Reference drifts

Voltage references come in many initial tolerances and temperature drifts. The available temperature drifts range from as low as 3 ppm/°C for a good-quality reference to approximately 100 ppm/°C for a low-cost version. The drifts can be either positive or negative with respect to changes in temperature depending on the type of reference chosen. Once the temperature range is known, the reference can be chosen with the drift specification matching the application. An option for very high-precision systems was described in Chap. 3. This automatic technique first performed a "signature" measurement of the reference over the operating range for subsequent corrections. The following example shows the amount of error caused by a typical reference with 20 ppm/°C and an initial temperature of 25°C.

Conditions:

$$V_{ref} = 10.000\text{ V} \qquad \text{"perfect" at } 25°\text{C}$$

$$\text{Temperature range} = -40 \text{ to } +85°\text{C}$$

$$\text{Error} = 10\text{ V } [20/1\text{M(C)} \times (85 - 25)°\text{C} = 12\text{ mV}]$$

$$\%\text{ error} = 100\left(\frac{12\text{ mV}}{10\text{ V}}\right) = 0.12 \qquad (0.1\% = 10\text{-bit accuracy})$$

Always pay attention to the temperature drift of the reference so that you do not pay for high initial accuracy at 25°C while the drift greatly exceeds it.

Circuit board thermal effects

The thermocouple sensor is based on the fact that two dissimilar metals produce a voltage change proportional to a temperature variation. Unfortunately, this same principle applies to the components in a circuit board. If there are any temperature gradients present within the precision analog circuitry, the potential thermal drifts could prove unacceptable. For instance, parasitic thermocouples exist between a copper circuit board and integrated packages with lead frames made of either copper or nickel-alloy material. Each junction soldered to the circuit board will generate a voltage as a function of temperature approximately equal to

Copper leads:	1 to 3 μV/°C
Nickel-alloy:	18 μV/°C

Precision resistors (i.e., metal-film) are usually made with copper leads and solder. This reduces the temperature drift to about 0.1 to 0.2 μV/°C. Resistors are generally not a problem since the two leads will generate equal but opposite voltages and therefore will cancel. If a temperature gradient does exist, it is possible to use two resistors in series (even if only one is required), as shown in Fig. 9.9. With this technique, the sets of junctions will be very close together, thus making the two soldered lead temperatures the same. This effectively cancels the thermally generated voltages.

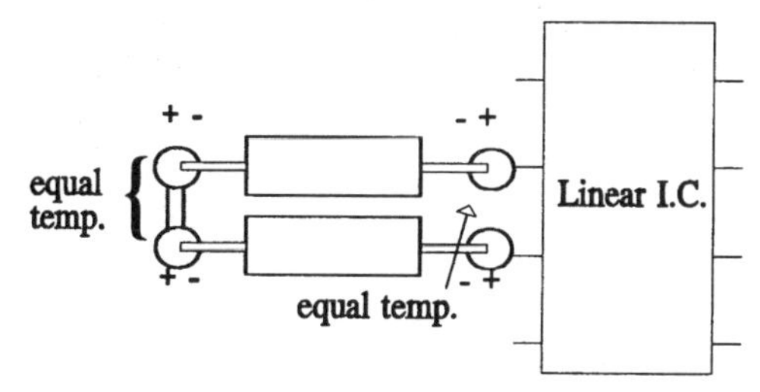

Figure 9.9 Resistor connection to avoid thermal gradient problems.

Guidelines for avoiding thermoelectric effects:

1. Keep temperature gradients to a minimum, if at all possible. Separate power from precision components, and put precision components as close together as possible.
2. Use an even number of connections of the same material to cancel thermoelectric effects.
3. Use shielding to prevent forced air over precision circuitry.

Low-Level Currents

To keep the circuit board leakage less than the low-input-current CMOS or BiFET amplifier input leakages requires special attention. Generally speaking, a low-level current is considered anything less than 10 nA. Standard epoxy circuit boards can create leakage paths due to voltage differentials between adjacent traces. The most likely leakage paths are from the amplifier supply voltage to the input pins. Even with a circuit insulation resistance of $10^{12}\,\Omega$, a potential difference of 15 V can cause a leakage current of 15 pA. This may not seem like much, but it is when a CMOS amplifier with less than 1 pA of input leakage current is required. Also keep in mind that circuit board leakage will be much worse when the board has not been cleaned properly or is subjected to high humidity during operation.

When leakage problems do occur, they often cause strange results that are hard to figure out. One reason is the difficulty in measuring the various circuit test points without introducing more leakage paths. Teflon-coated (as opposed to plastic-coated) probes are the best for minimizing leakage during test measurements. Leakage currents will be dramatically reduced by using guard rings to protect the input pins from circuit board leakage effects. Guard rings are simply copper rings around the input terminals of the amplifier (Fig. 9.10*a*) on both sides of the circuit board. It is important to connect the ring to a low-impedance voltage of the same potential as

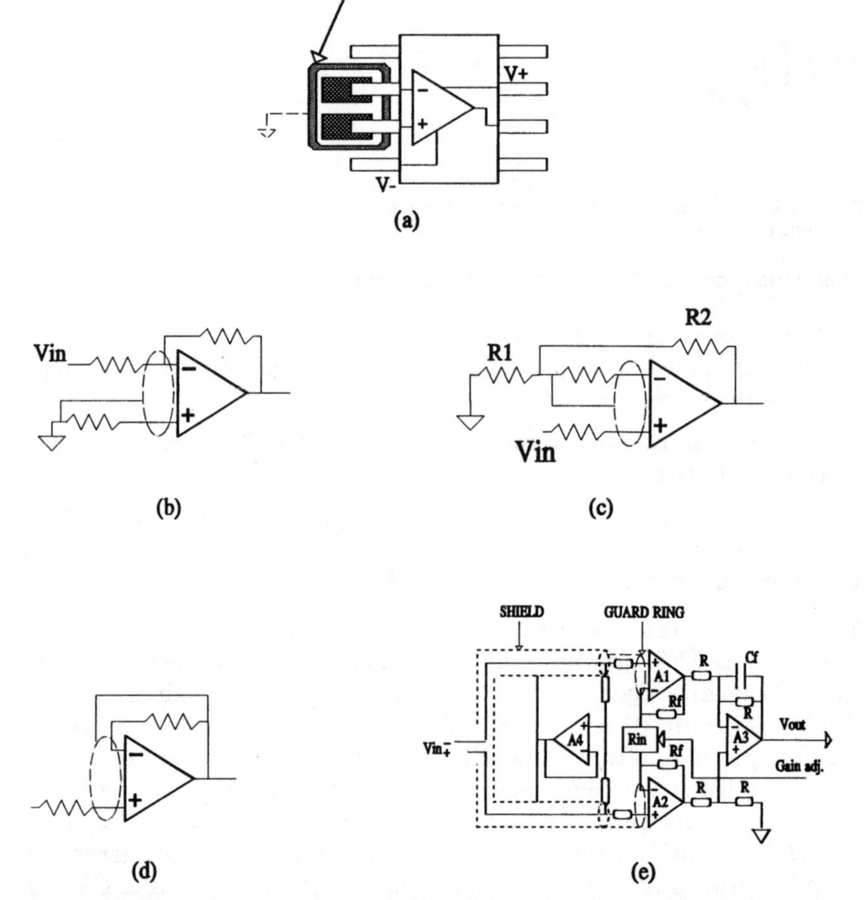

Figure 9.10 Examples of amplifier guard rings. (*a*) Guard ring example for 8-pin DIP/S.O. pkg.; (*b*) inverted amplifier; (*c*) noninverting amplifier; (*d*) buffer amplifier; (*e*) instrumentation amplifier.

the input pins. This effectively isolates the amplifier pins from the circuit board. Figure 9.10*b* to *e* illustrates how to create guard rings for various amplifier configurations. The instrumentation amplifier in Fig. 9.10*e* illustrates the technique of driving the guard ring with an amplifier. This is the same circuit that was used in Chap. 5 (Fig. 5.14) for the strain gage example. When a shield is used as in this application, the guard ring should be connected directly to it.

In some cases, the circuit board leakage will be a problem regardless of what you do, and it may be necessary to use the stand-up technique (air is an excellent insulator). For example, instead of connecting the input resistor and capacitor of an integrating amplifier through the circuit board, solder the components directly to the amplifier input pin. It is also not a good idea to use a socket since all this will do is potentially trap contaminates or provide another leakage path. If you have to use one, a Teflon socket is recommended. To avoid problems after the design is in service, thoroughly clean, bake, and use a conformal coating to isolate sensitive circuitry from moisture and other contaminants.

Good Layout Practices

Experienced analog designers realize that circuit board layout is extremely important to the performance of the design. Great care during layout will avert problems later from several sources. The process can become complicated when you're trying to incorporate precision analog, high-frequency digital, and power electronics together on one board. Fortunately, there are several design rules that, if followed, should make it much easier to achieve the desired performance from the system.

When a circuit board layout is started, the whole circuit must be kept in mind. This will facilitate proper partitioning between the various functions. For instance, the precision analog section should be located close to the input signals and separated from digital and power electronics. The objective here is to isolate the potentially noisy signals and to facilitate proper routing of the ground, power, and signal connections. With appropriate partitioning, digital and power lines are prevented from crossing the analog section, and circuit board traces will be minimized.

It is common practice for A/D converters to provide separate analog and digital power/ground pins. How you connect these pins externally will depend on whether there are multiple ADCs in the design. It is recommended that single-ADC systems tie the analog and digital pins together by connecting the analog and digital ground planes at the chip. Multiple A/D converters should run the analog and digital grounds separately back to the power supply. Either way, only one common connection should be made between the analog and digital grounds. Thus, it prevents ground loops which are voltage drops caused by currents across circuit traces. Figure 9.11 illustrates a typical layout for a single ADC.

Note that there are separate ground planes for the analog and digital portions of the ADC, and the two ground planes are connected at the chip. Ground planes serve important functions: First, maximum noise attenuation is achieved by immediately routing unwanted signals to ground through proper paths. Second, analog and digital power and reference supply pins are efficiently decoupled to their correct ground through minimum inductance and resistance. Standard procedure for decoupling is to use a ceramic in parallel with a large

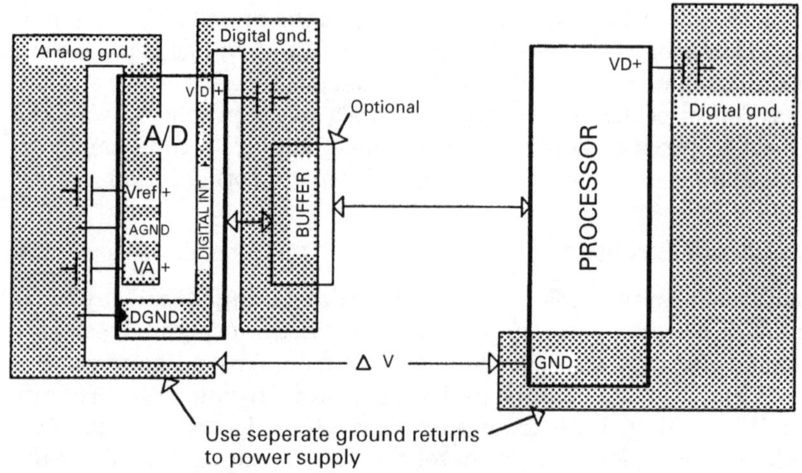

Figure 9.11 Basic ADC layout with ground planes.

(i.e., more than 10 μF) tantalum or electrolytic capacitor. It is absolutely critical that the lead lengths for the high-frequency ceramic capacitors be as short as possible, to minimize inductance. Chip capacitors offer the best way of achieving minimum inductance. One final point: A digital buffer can be placed between the ADC and the system processor to further isolate digital noise from analog circuitry (shown as optional in Fig. 9.11).

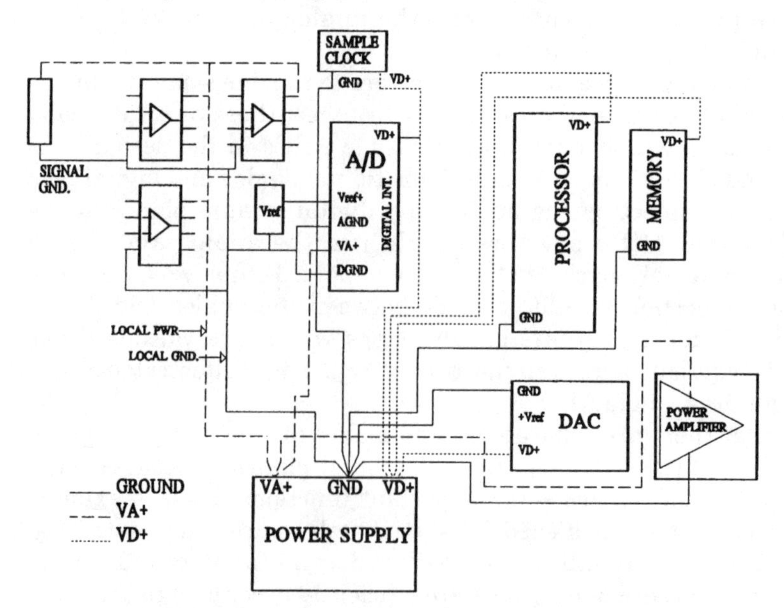

Figure 9.12 System partitioning for proper layout of ground, power, and signals.

System power and ground routing

Although there are infinite variations in mixed analog and digital systems, several guidelines for routing the power to each section are applicable. The fundamental objective is to avoid ground loops, minimize noise generation, and minimize coupling between analog and digital sections. Ground loops can be created when two or more circuits are connected in series. Operating current from these devices will result in voltage differences between the individual ground levels. Many analog and power application notes recommend using a single-point ground connection to avoid ground loops. Figure 9.12 illustrates the layout strategy for a "typical" mixed signal system.

Ideally, each integrated circuit (IC) should have a separate ground connection back to the power supply input ground. However, this may not be practical when there are several ICs in the system. One option is to use local ground planes to connect common low-power circuitry and to run a connection back to the power supply ground, using as wide and short a trace as possible.

Power supply connections should also be routed depending on the function and power levels. For instance, when more than one device uses a common supply such as a power amplifier and the signal conditioning amplifiers, a separate power bus should be used for each. This is similar to the single-point ground method mentioned above, with power lines making a common connection at the main power supply source. Whenever possible, use separate power supplies for the analog and digital sections, and always use decoupling capacitors for each chip.

Summary

1. Isolate sensitive analog circuitry from digital and power devices by proper partitioning of the system.
2. Minimize circuit traces (minimize noise pickup and voltage drops).
3. Use single-point ground connections.
4. Use analog and digital ground planes for the return of unwanted high-frequency signals, and make only one connection between them.
5. Use guard rings and shielding for low-level amplifier circuitry.
6. Avoid crossing analog sections with digital traces (if necessary, use right angles).
7. Connect signal ground to analog ground to avoid voltage offsets.
8. Liberally use decoupling capacitors with short lead connections for each IC.

9. Isolate the A/D sampling clock from the high-speed digital circuits to limit clock jitter.

Autocalibration Techniques

System designs that operate at very high speed or use a sensor (or other components) with low initial tolerance often require some form of autocalibration to maintain precision. Common in all forms of autocalibration is that the processor will read the input signal, make corrections until the error is acceptable, and then store in memory the required compensation. An example of a system that typically requires autocalibration is the digital storage oscilloscope. In this application, compensation can be made for high-speed buffers that typically do not have good dc performance. Another example that requires the autozero function is a weighing scale. When an empty container is placed on the load cell platform, first the scale must be zeroed out before the actual material (called the *tare* weight) is weighed.

Autocalibration can consist of autozero and/or autogain depending on the application. Considerations when you are performing an autocalibration operation include the level of hardware or software required and whether the system should be autocalibrated periodically or initiated by the operator. An important point is that calibration should be performed only after the system has been allowed to cool down after soldering and warm up sufficiently after power is applied. Otherwise, the calibration may not hold the desired accuracy.

Autozero correction

One autozero method is to use a D/A converter as shown in Fig. 9.13. Under processor control with a "zero input," the D/A converter provides an output ranging from $+V_{ref}$ to $-V_{ref}$ for subtracting (or zeroing) with the input amplifier stage until the ADC output reads zero. This method is useful for weighing applications to null unwanted weight (i.e., container) while maintaining the full accuracy over a wide input range. By reducing V_{ref} or lowering the gain of the circuit in Fig. 9.13, calibration in very small increments can be made for low-level circuit offsets.

$$V_{out} = \frac{V_{ref}(D - N')}{N'}$$

where D = digital input and N' = resolution/2 (that is, N' = 512 for 10-bit resolution).

If autozero is required only to zero out small offsets from the circuit, then a simpler method is possible. Again, with processor assistance, a ground input can be selected with an analog switch instead of using a DAC approach (Fig. 9.14*a*). After a measurement is taken, the

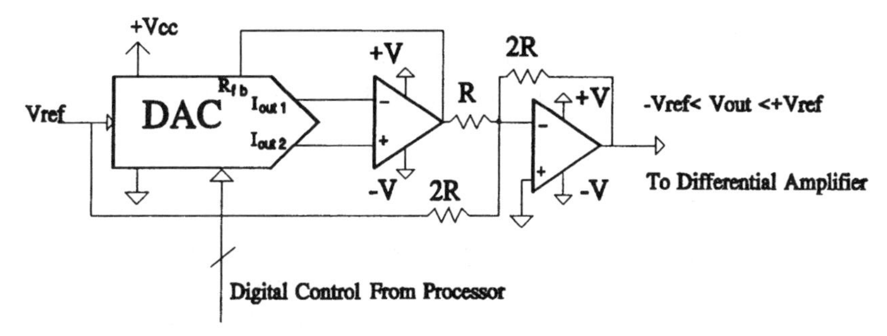

Figure 9.13 Autozero using a D/A converter.

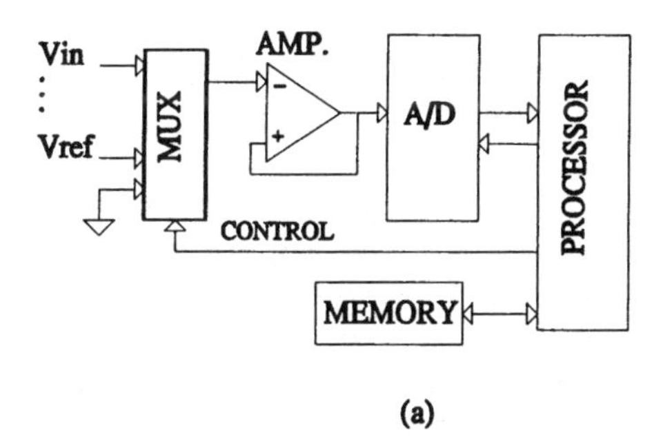

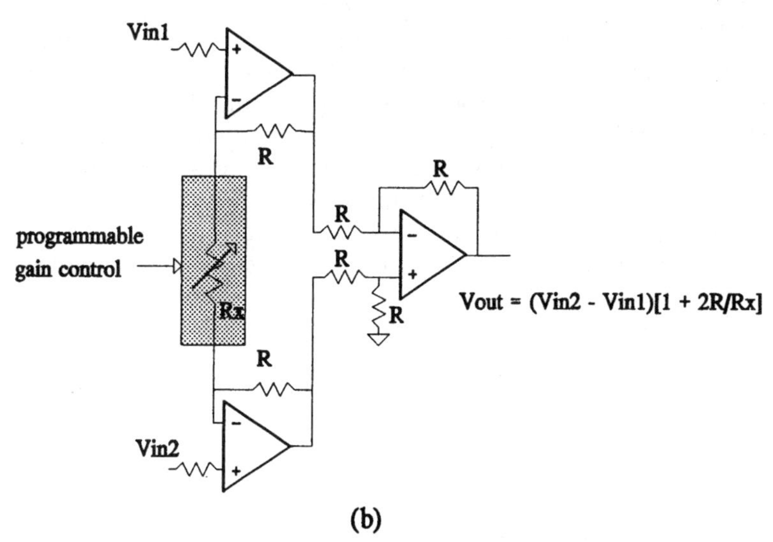

Figure 9.14 Examples of autogain control. (*a*) Software gain correction with V_{ref} input; (*b*) digital gain control with analog switch.

result is stored in memory (either RAM or EEPROM) for subtraction from each subsequent measurement. Software overhead to perform the subtraction is very low and should not pose a problem with most applications.

Autogain correction

Autogain correction is useful when either the data acquisition system must measure a wide range of inputs or the circuit gain does not maintain the desired accuracy. How this is done will depend on the amount of gain adjustment required. Fine gain adjustments for circuit error compensation can be performed with software by switching in V_{ref} and then storing the full-scale measurement in memory, as shown in Fig. 9.14*a*. However, there is a drawback to this approach. Each input measurement must be adjusted by performing a multiplication and division operation, and in some applications, this can cause a throughput problem.

There are several options for providing digital gain adjustment with the use of analog switches, as shown in Fig. 9.14*b*. Either standard analog switches or special control switches such as the National Semiconductor LF13006/7 digital gain control switches are an option, with the LF13006 providing gains of 1, 2, 4, 8, 16, 32, 64, and 128 and the LF133007 providing gains of 1, 2, 5, 10, 20, 50, and 100. This technique can provide either narrow (fine) adjustments to compensate for circuit tolerances or wide (course) adjustments for accommodating large input signal ranges. If circuit tolerances are not an issue and the input range is wide, then a fully integrated programmable gain amplifier (PGA) is another option for course gain range adjustments.

Still another option is to use an ADC with very high resolution, and not use any gain adjustment. For instance, providing the circuit tolerances are acceptable, wide input signal ranges can be accommodated by using a 16-bit ADC even though the required accuracy may be only 10 bits. Thus, 10-bit accuracy is maintained with a full-scale input range varying by as much as 6 bits (64). Potentially this can be less expensive overall and can avoid the problems with errors introduced by the extra hardware.

Bibliography

1. *Analog-Digital Conversion Notes,* Analog Devices, Norwood, MA, 1977.
2. Candy, James C., and Temes, Gabor C.: *Oversampling Delta-Sigma Data Converters,* IEEE Press, Piscataway NJ, 1990.
3 Stout, David F., and Kaufman, Milton: *Operational Amplifier Circuit Design,* McGraw-Hill, New York, 1976.
4. Frederiksen, Thomas M.: *Intuitive Operational Amplifiers,* McGraw-Hill, New York, 1988.
5. High-Speed Design Seminar, Analog Devices, Norwood, MA, 1989.
6. Seales, Alan: *Linear and Conversion Application Handbook,* AN-15, "Minimization of Noise in Operational Amplifier Applications," Precision Monolithics Inc., 1988.
7. Givens, Shelby D.: *Linear and Conversion Application Handbook,* AN-35, "Understanding Crosstalk in Analog Multiplexers," Precision Monolithics Inc., 1988.
8. Louzon, Paul: "Using Harris High Speed A/D Converters," AN9214.1, 1993.
9. Witte, John: "Reducing CMOS Flash Converter Dissipation," *Electronic Engineering,* November 1989, Datel, pp. 22, 26.
10. Kester, Walt: "Flash ADCs Provide the Basis for High-Speed Conversion," *Electronic Design News,* January 1990, pp. 101–110.
11. Daugherty, Kevin M.: "Try PWM for Low-Cost A/D Conversion," *Electronic Design,* January 1990, pp. 51–54.
12. Daugherty, Kevin M.: "Try Single-Slope A-D Conversion for a Low-Cost 12-Bit Solution," *Electronic Design,* January 1992, pp. 59–66.
13. Sylvan, John: "Build Precise S/H Amps for Fast 12-Bit ADCs," *Electronic Design,* January 1990, pp. 57–62.
14. Angello, Anthony: "16-Bit Conversion Paves the Way to High-Quality Audio for PCs," *Electronic Design,* July 1990, pp. 61–66.
15. Goodenough, Frank: "20-Bit Delta-Sigma ADCs Vie for Integrator Jobs," *Electronic Design,* April 1991, pp. 93–96.
16. Demler, Michael J.: *High-Speed Analog-to-Digital Conversion,* Academic Press, New York, 1991.
17. Sockolov, Steve, and Wong, James: "High-Accuracy Analog Needs More Than Op Amps," *Electronic Design,* October 1992, p. 57.
18. Kester, Walt: "Layout, Grounding, and Filtering Complete Sampling ADC System," *Electronic Design News,* October 1992, pp. 127–134.
19. Koljonen, Tuomas, and Vuori, Jarkko: "Averaging Increases μPs ADC Resolution," *Electronic Design News,* February 1990, pp. 139, 140.
20. Kester, Walt: "Peripheral Circuits Can Make or Break Sampling ADC Systems," *Electronic Design News,* October 1992, pp. 97–106.
21. Jung, Walt: "Build an Ultra-Low-Noise Voltage Reference," *Electronic Design,* June 1993, pp. 74, 75.
22. Bodio, Mark: "Reference Compensation Minimizes Temperature-Related Problems," *Electronic Design News,* February 1993, pp. 135–140.

23. Rahim, Zahid: "Choosing the Right Sample-and-Hold Amplifier," *Electronic Products,* December 1991, pp. 135–138.
24. *Delta-Sigma Techniques,* AN 10REV1, Crystal Semiconductor Corp., January 1989, Austin, TX.
25. Rappaport, Andy: "Capacitors," *Electronic Design News,* October 1982, pp. 105–122.
26. Yates, Warren: "Designer's Guide to Film Capacitors," *Electronic Products,* September 1982, pp. 105, 106.
27. Conner, Doug: "Selecting Capacitors for High Density Circuit Applications," *Electronic Design News,* December 1990, pp. 80–86.
28. Kerridge, Brian: "Elegant Architectures Yield Precision Resistors," *Electronic Design News,* July 1992, pp. 86–92.
29. Ormond, Tom: "Resistor Networks," *Electronic Design News,* June 1989, pp. 125–132.

Index

ABOUT THE AUTHOR

Kevin M. Daugherty is an automotive technical manager with Temic, the semiconductor division of Daimler-Benz, where he develops new automotive product strategies and provides technical assistance to customers. He previously worked at National Semiconductor and Bell Northern Research. Mr. Daugherty received a patent on his variable range PWM analog-to-digital technique, and has written related articles for *Electronic Design* magazine. During his seven years at National Semiconductor as a field application engineer, he worked to solve customers' analog-to-digital design problems. He received his bachelor's degree in electronic engineering from Wayne State University, and is a member of the Electronic Engineering Honor Society.